THE GLACIATION OF THE ECUADORIAN ANDES

Chimborazo from South

THE GLACIATION
OF THE
ECUADORIAN ANDES

STEFAN HASTENRATH

University of Wisconsin, Madison

CRC Press
Taylor & Francis Group
Boca Raton London New York

CRC Press is an imprint of the
Taylor & Francis Group, an **informa** business
A BALKEMA BOOK

Published by:
CRC Press/Balkema
P.O. Box 447, 2300 AK Leiden, The Netherlands
e-mail: Pub.NL@taylorandfrancis.com
www.crcpress.com – www.taylorandfrancis.com

© 1981 by Taylor & Francis Group, LLC
CRC Press/Balkema is an imprint of the Taylor & Francis Group, an informa business

No claim to original U.S. Government works

ISBN 13: 978-90-6191-038-1 (hbk)

Visit the Taylor & Francis Web site at
http://www.taylorandfrancis.com

and the CRC Press Web site at
http://www.crcpress.com

To my mother and
the memory of my father

CONTENTS

ILLUSTRATIONS AND TABLES

FIGURES

PHOTOGRAPHS

MAPS

TABLES

XI

ACKNOWLEDGEMENTS

This inevitably incomplete study would not have been possible at all without the co-operation and support I received in the course of the years. The three field seasons, December 1974-January 1975, May-June 1975 and May 1978, were generously supported by the Dirección General de Geología y Minas (DGGM), Ecuador. Ing. Salvador Guevara accompanied me during the first, Ing. Wilson Santamaría in the second, and Ing. Manuel Cholango in the third season. The 1978 field work was also subsidized by the Nave Bequest Fund through a grant from the Research Committee of the University of Wisconsin Graduate School, Madison. The evaluation of results was in part supported through National Science Foundation Grant EAR76-18881. The Instituto Geográfico Militar in Quito placed ozalid copies of unpublished topographic sheets at my disposal. Lic. Alejandro Reinoso and his staff assisted me in the search for air photographs. Mrs Eva Singer at the University of Wisconsin typed the various generations of the manuscript. I am grateful to Prof. Thomas van der Hammen, Universiteit van Amsterdam, for contributing Appendix IV; and to Borntraeger Publishers for making Fig. 15 available.

Prof. Karl W. Butzer, University of Chicago, Prof. Minard L. Hall, Escuela Politécnica Nacional-Quito, Dr Carlos Schubert, IVIC-Caracas, Dr Ernst Loeffler, CSIRO-Canberra, kindly read draft versions or portions of the manuscript. I am indebted to Prof. Minard L. Hall, for numerous discussions and an illuminating field trip. I further acknowledge discussions with Ings. Gaston Ruales M., Rodrigo Alvarado B., Jorge Guzmán, F. Mosquera, Drs Humberto Sosa, Brian J. Kennerley, Roger C. Bristow, Alan Wilkinson, R.B. Randel, and C. Mortimer, of the DGGM; with Ings. Eduardo Mancheno, Vicente Lauro Gómez A., Capt. Carlos Blandín, of the Instituto Nacional de Meteorología e Hidrología e Hidrología, Quito; and with Dr Emilio Bonifaz, Quito. I thank Dr Jorge Lara and Capt. Gustavo Gomez Jurado, Quito, for making a survey flight possible.

I enjoyed the hospitality and companionship of the mountaineering community of Quito: Club de Andinismo Politécnico, Nuevos Horizontes, International Andean Mountaineers, Cumbres Andinas, Club de Andinismo del

Colegio San Gabriel, El Sadday. Conversations with Martin Slater, Manuel and Francisco Rivas, Dr Jorge Montalvo, Ings. Pablo Andrade, Santiago Rivadaneira, Adolfo Holguín, Bernardo Beate, Miguel Astudillo, were particularly informative concerning the ice extent in remote mountain areas. Finally, my thanks go to Marco Cruz Arellano, who best knows the Ecuadorian Andes.

Quelccaya, July 1979   Stefan Hastenrath

1 INTRODUCTION

... deux chaînes parallèles de hautes montagnes, qui
font partie de la Cordelière des Andes. Leurs cimes se
perdent dans les nues, & presque toutes sont couvertes
de masses énormes d'une neige aussi ancienne que le
monde. De plusieurs de ces sommets, en partie écroulés,
on voit sortir encore des tourbillons de fumée & de
flamme au sein même de la neige.

Charles Marie de La Condamine, *Journal du voyage a
l'Équateur*, 1751.

Hochgebirgsforschung versprach gerade dort sehr
interessante Ergebnisse, weil vielleicht kein andres Land
der Welt eine solche Fülle von natürlichen Gegensätzen
in sich vereint ... wie das ... aus tropisch-heissen
Niederungen zu den Regionen ewigen Schnees aufge-
türmte Andenland Ecuador.

Hans Meyer, *In den Hochanden von Ecuador*, 1907.

The High Andes of Ecuador are among the three regions of the World
where glaciers still exist in immediate vicinity of the Equator. The area
covered by ice caps and glaciers is only a small fraction of the total land
surface and their influence on the regional climate may be of subordinate
importance. However, glaciers in terms of their response to atmospheric
forcing are extremely sensitive — albeit complex — indicators of large-scale
environmental change. Accordingly, the assessment of long-term glacier
behaviour has been recognized as a task of high priority in international
efforts at climatic change (UNESCO 1970; World Meteorological Organization — ICSU 1975:7, 11, 16; International Association of Hydrological

1

Sciences – UNESCO 1977; Temporary Technical Secretariat for World Glacier Inventory, UNESCO-UNEP-IUGG-IASH-ICSI 1977). In view of the drastic ice retreat since the latter part of the past century borne out for the other two glaciated high mountain regions under the Equator – New Guinea (Hope et al. 1976) and East Africa (Hastenrath 1975) – it appears timely to reassess the glaciation of the Ecuadorian Andes.

The present study attempts to assemble an inventory of the current ice extent; to reconstruct glacier variations in the course of the past several centuries; and to infer glacial-climatic events in the more distant geological past. The time has become ripe for such an undertaking in the course of very recent years, with the systematic aerial photography and topographic mapping by the Instituto Geográfico Militar of large parts of the High Andes (Appendix I). This background was essential for the field survey of recent glaciation and fossil glacial morphology conducted during a preparatory travel in June 1969 and three field seasons, December 1974-January 1975, May-June 1975 and May 1978. A survey flight of Chimborazo and El Altar became possible during the latter season. Short information visits materialized in June 1977 and in June 1979. Topographic maps and air photographs also permitted a limited evaluation of current ice extent and moraine morphology for mountain regions that could not be visited during the field work. Field survey and air photo interpretation were combined with an evaluation of a variety of historical sources, such as journals, sketches and photographs. The investigation was further complemented by enquiries from mountaineers familiar with the more remote regions.

Variations in the recent glaciation can in part be reconstructed from land descriptions in the early era of Spanish colonization, and from the accounts of expeditions to the Ecuadorian Andes in the course of the past more than 200 years (Appendix II). The earliest usable references date from the 16th century and are contained in the municipal and ecclesiastic archives of Quito, and the Spanish Archivo de Indias (Rodriguez de Aguayo 1965). Although the Quito municipal and church records from 1534 to 1576 and 1593 to 1657 are available in published form, except for shorter breaks (Gonzalez Rumazo 1934a, b; Garcés 1934a, b, 1935, 1937-40, 1941a, b, 1944, 1946-47, 1955, 1960; Schottelius 1935-36; Chirriboga 1969; Bonifaz 1971), no information bearing on glacier variations could be found for the 150 year gap from the 1580's to the 1730's. Subsequently, important references are due to the geodetic mission of the French Academicians in 1735-43 (Juan & Ulloa 1748; Bouguer 1749; La Condamine 1751); and to Velasco (1841-44), who lived in Ecuador from his birth around 1727 until his departure for Europe in 1767. A. von Humboldt's (1810, 1853, 1874a, b) travels led through Ecuador in 1802; his observations of Nature are particularly perceptive and careful. Boussingault (1935a, b, 1841, 1849, 1851) visited the country in 1831. In 1858 Villavicencio published the first textbook on the geography of Ecuador. G. García Moreno and S. Wisse wandered through the Sierra extensively around the middle of the 19th century (Wisse & García Moreno 1846; Wisse 1849; Loor 1953). R. Spruce (1861) travelled through Ecuador in 1857, M. Wagner (1870) in 1858-59, W. Jameson (1861) in 1859 and J. Orton (1868, 1869, 1870) in 1867.

The German geologists W. Reiss and A. Stübel carried out extensive research in the course of 1870-74. Their objectives were mainly concerned with volcanism, but their accounts include careful observations on the recent glaciation. In particular, they were accompanied by the Ecuadorian painter R. Troya, who produced 64 paintings of Andean landscapes. These paintings were exhibited at the Grassi Museum in Leipzig; however, according to recent information this collection does not seem to have survived World War II. The results of Reiss and Stübel's expedition were published with great delay (Reiss 1873, 1875; Reiss & Stübel 1892-98; Stübel 1886, 1897; Dietzel 1921). Kolberg (1885) and Dressel (1877, 1879, 1880) travelled in Ecuador in the 1870's. The renowned first ascents of E. Whymper (1892) took place in 1880. Although his remarks on glacial phenomena lack precision, his book contains several etchings of peaks that provide useful historical documentation. T. Wolf (1878, 1892) travelled extensively in Ecuador, and in 1892 he published another textbook on the regional geography and geology. P. Grosser (1905) pursued volcanological studies in 1902. H. Meyer's (1904a, b, c, 1906-7, 1907) expedition in 1903 is a further milestone in the exploration of the Ecuadorian Andes in that his research focussed on both the fossil and recent glaciations. Some of his results have been summarized by Mercer (1967). Meyer's photographs, sketches, and descriptions represent important historical documentation for the reconstruction of secular glacier variations.

Photographic documentation since then is sporadic. N. Martinez (1933), the pioneer in Ecuadorian mountaineering, took numerous photographs mostly around 1906, which have been preserved (archive of 'Nuevos Horizontes', Quito). Occasional photographs from intermediate decades were obtained through the mountaineering community in Quito. And a fine collection of pictures is contained in Eichler's (1952) book. Further photographs of interest are found in Andrade Marín (1936), Blomberg (1952), and Sauer (1971).

Theories concerning glaciation in the more distant geological past date back to the past century. Reiss (Reiss & Stübel 1892-98:162-165) did not fail to recognize the wealth of moraine morphology attesting to a formerly much larger ice extent in the High Andes of Ecuador. However, he viewed the ice retreat as resulting from the gradual lowering of mountain peaks through erosion. He categorically ruled out glacial episodes of a climatic nature in Ecuador.

Meyer (1907:175-177, 273, 364-368, 395-397, 445-467) was aware of

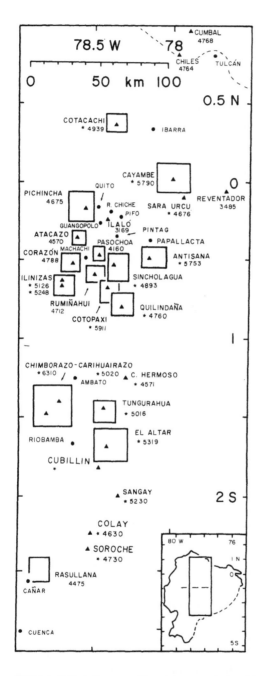

Figure 1. Orientation map. Mountain areas depicted in Maps 1-16 are marked by boxes. Asterisks denote mountains presently glaciated. Approximate location of transects in Figures 2 and 11 is indicated by broken line.

former glaciations in the tropics from his experiences at Kilimanjaro. He proposed two glacial-pluvial periods equivalent to the European Riss and Würm, separated by a dry interglacial. He ascribed the ubiquitous moraines extending down to around 3700-3800 m to the younger of the two glacials, and inferred an ice retreat in three phases from triple moraine ridges. Rock ledges at the sides of glacial valleys and two sets of large river terraces are adduced as evidence for the older, more pronounced glacial.

Sauer (1950, 1965:260-288, 1971:145-165) proposed four glacials including one oldest, pluvio-glacial period, separated by three interglacials, but offered no correlation with Meyer's (1907) scheme. Interference of epirogenic uplift and sinking with climatic episodes is considered in particular. Sauer ascribed a voluminous collection of faunal remnants to the third interglacial; and hardened spherical configurations within layers of weathered tuff, Cangagua, are described as reference horizons for the second and third interglacials.

Sauer deduced his scheme primarily from stratigraphy in the profile of rivers, rather than moraine morphology in the high regions. Glacier striations in the Quito region, on the slope of the Eastern Cordillera, are ascribed to the second glaciation. For the second glaciation a limit as low as 1600 m is claimed in the Western Cordillera, and the third glaciation is regarded as rather more extensive. The fourth glaciation is described as weaker, but in part more extended than the third glaciation (Sauer 1971:156). This is offered as explanation for the absence of moraines of the third, and the preservation of moraines of the fourth glaciation. According to Sauer (1971: 158), mountains as low as 3100 m would have been the origin of large glaciers during the fourth glaciation. Sauer (1971:153) ascribed three terrace levels in the valleys of the cordilleras to the second, third and fourth glaciations.

It does not seem possible to reconcile the schemes of Meyer and Sauer. In terms of preservation of moraine morphology, at best Sauer's fourth glaciation could be matched to Meyer's younger, 'Würm', glaciation; however, the glacier limits of 3700-3800 m of Meyer are well above the snow line elevation of less than 3100 m claimed by Sauer. Because of the absence of moraine morphology, the first through third glaciations of Sauer would have to predate Meyer's younger, 'Würm', and older 'Riss' glacials, somewhat at variance with Sauer's triple terraces ascribed to the latter three glaciations of his scheme.

After a preparatory review of the gross physiographic structure (Chapter 2) and large-scale circulation and climate (Chapter 3), evidence on both the recent and fossil glaciations shall be presented in Chapters 4 to 6 separately for each mountain (Figure 1). Topographic charts and air photographs are listed in Appendix I, and historical sources are given in Appendix II. A synthesis of glacial chronology for the more distant geological past, and secular variations of the modern glaciation shall be attempted in Chapter 8.

5

2 PHYSIOGRAPHIC STRUCTURE

Diese Vulkanberge sind trotz ihrer grossen Zahl
orographische sehr übersichtlich angeordnet; sie grup-
piren sich vornehmlich in zwei sich meridional er-
streckende Ketten, welche das Hochland Ecuadors an
seinen Rändern, im Osten und im Westen, einfassen;
man spricht deshalb kurzweg von der Ost- und West-
cordillere; ihre inneren Seiten sind einander zugekehrt
und beherrschen das Hochplateau; ihre äusseren fallen
zur Tiefe: die Ostcordillere gegen das Amazonasgebiet,
die Westcordillere gegen das Stille Meer.

Alphons Stübel, *Die Vulkanberge von Ecuador*, 1897.

In the equatorial region, the Andes are compressed as a narrow band between
the Brazilian shield and the Pacific mass (Lewis 1956:270). Three major
physiographic provinces are distinguished in mainland Ecuador (Gerth 1939,
1955; Tschopp 1948; Lewis 1956; Sauer 1957, 1965, 1971; Dirección
General de Geología, Minas y Petróleo 1969): the coastal plains in the West
(Costa), the Andean highlands (Sierra), and the descent to the Amazonian
lowlands in the East (Oriente). A schematic zonal transect across these essentially North-South trending entities is shown in Figure 2.

The Andean province is made up of three meridionally oriented divisions,
namely the Western Cordillera (C. Occidental), the Interandean Depression
(Depresión Interandina), and the Eastern Cordillera (C. Oriental or Real).
The Interandean Depression represents a huge graben bounded by fault
zones. It consists of several basins draining alternately towards the Amazonian lowlands and the Pacific coast. The Eastern Cordillera of Ecuador
represents the southward extension of the Colombian Central Cordillera.
Corollaries of the Eastern Cordillera of Colombia are found in the Cordilleras Cutucú and Condor in the Oriente of Ecuador, and were termed 'La
Tercera Cordillera' by Sauer (1957:12; 1965:18). The Western Cordillera

6

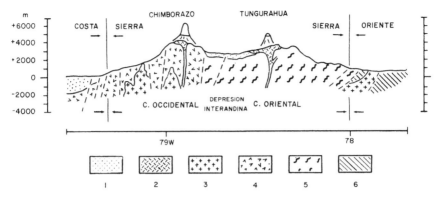

Figure 2. Schematic zonal transect across Equatorial Andes at approximately 1°S.
1. Quaternary sediments; 2. recent volcanic rocks; 3. igneous rocks; 4. cretaceous volcanic (Piñón) and cretaceous to paleocenic sedimentary (Flysch, Yunguilla San Marcos) rocks; 5. presumed carboniferous or older metamorphic and semi-metamorphic rocks (Margajitas, Gualaquiza, Punta Piedra); 6. Jurassic and cretaceous sedimentary rocks (Napo, Chapiza-Misahualli). Vertical exaggeration of topography is fivefold, as in Figure 11. Strongly simplified after Servicio Nacional de Geología y Minería (1969).

of Ecuador lacks an immediate connection with the Cordillera Occidental of Colombia (Sauer 1971:259).

The high volcanoes are aligned mainly along the great fault zones on the West flank of the Cordillera Oriental and the East flank of the Cordillera Occidental. However, the volcanoes El Reventador and Sangay rise further East near fault zones in the Cordillera Oriental. And a few smaller volcanoes, such as Rumiñahui, sit in the Interandean Depression itself, presumably over subordinate fault zones (Lewis 1956:274). Of the presently snow-capped peaks shown in Figure 1, few are not of volcanic origin: the Cerro Hermoso in the Llanganates and the Sara Urcu are made up of metamorphic rocks pertaining to the palaeozoic base of the Andean orogene.

The Andean orogene is overlain by quaternary deposits to the West; and limestones and clays of the cretaceous Napo formation, and pyroclastics and sandstones of the jurassic-cretaceous Chapiza-Misahualli formation, bound it along fault zones in the East. The basis of the Andean orogene is in the East mainly made up of metamorphic rocks of probably cambrian age or older; in the West it consists in large part of diabases, porphyrites, and pyroclastics of the cretaceous Piñón formation, and a variety of igneous rocks. The Interandean Depression and the adjacent cordilleras carry thick layers of volcanic ashes. Some of the older deposits are known by the name 'Cangagua' (Sauer 1966, 1971). Quaternary volcanism, largely of andesitic nature, is concentrated in the Western and Eastern Cordilleras. It is here that gigantic volcanic peaks were built up — that were to become the seat of spectacular ice caps and glaciers.

7

3 ATMOSPHERIC CIRCULATION AND CLIMATE

On pense ordinairement que pour passer du plus grand
chaud au plus extreme froid, il faut parcourir tout
l'intervalle qui separe la Zone torride des Zones froides;
au lieu qu'il suffit ici de monter mille ou douze cents
toises. Les climats les plus contraires s'y donnent pour
ainsi-dire la main; pendant que le haut est toujours cou-
vert de neige & que la neige y a en certains endroits
plus de 100 pieds d'épaisseur, on trouve en bas des Jar-
dins, & tous ces fruits qui ne viennent que dans les
endroits les plus chauds.

Pierre Bouguer, *La figure de la terre*, 1749.

Wohl wochenlang behält man von seinem Lagerplatze
aus einen wolkenbedeckten Berggipfel im Auge, um
seine Lage festzustellen oder um eine Zeichnung
gerade an dieser Stelle zu vervollständigen, sobald er
sich enthüllt. Aber vergeblich, er bleibt bedeckt! Und
fragt man die kundigen Begleiter, ob nicht bald eine
Besserung eintreten werde, so erhält man die Antwort:
'Así no más *vive* el cerro todo el año'.

Alphons Stübel, *Die Vulkanberge von Ecuador*, 1897.

The large-scale circulation over the Equatorial Americas (Figure 3) is in the
following discussed on the basis of surface and upper-air maps and meridional
and zonal vertical transects for the height of the Northern hemisphere winter
and summer seasons.

The surface climate of the sea areas surrounding the American Continents
has been studied recently on the basis of ship observations during 1911-70.
Data processing and analysis have been described in detail elsewhere (Hastenrath 1976a, b; Hastenrath & Lamb 1977; Hastenrath & Heller 1977; Covey

8

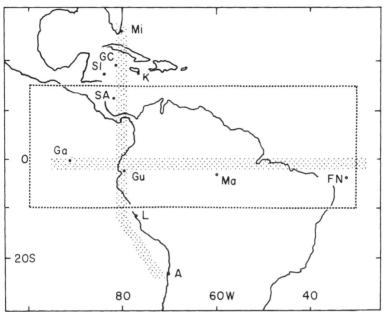

Figure 3. Orientation map. Area represented in Figures 4-6 is marked by broken line. Transects in Figures 7, 8, 12 are indicated by shading. Station identification: Mi – Miami, GC – Gran Caymán, K – Kingston, SI – Swan Island, SA – San Andrés, Gu – Guayaquil, L – Lima, A – Antofagasta, Ga – Galápagos, Ma – Manaus, FN – Fernando de Noronha.

& Hastenrath 1978; Hastenrath & Guetter 1978; Hastenrath 1978b). Figures 4 and 5 portray some characteristics of the surface wind field. Throughout the year (Figure 4) the Northeast trades sweep the North Atlantic, crossing the Central American land bridge into the Eastern Pacific. Here they meet the clockwise turning flow from the Southern hemisphere along a zonally oriented kinematic discontinuity. By contrast over the Western Atlantic, the trades from either hemisphere join in an asymptotic confluence zone. The near-equatorial confluence axis is farthest poleward during Northern summer, and reaches its equatormost position around February/March. Strongest surface convergence (Figure 5) does not coincide with the confluence axis (Figure 4): over the Eastern Pacific, maximum convergence is found to the South of the confluence axis throughout the year, with largest separation during Northern summer; in the Western Atlantic, the convergence maximum is to the North of the confluence axis in Norther winter, and to the South of it in summer. The convergence pattern in turn agrees closely with the distribution of cloudiness, precipitation frequency, and weather. Information of comparable quality is not available for the surface circulation over land.

A much coarser picture can be assembled of the upper-air flow patterns

9

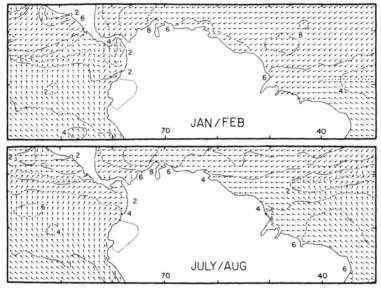

Figure 4. Surface resultant wind direction and speed (m s^{-1}) during January/February (top) and July/August (bottom). Position of near-equatorial confluence axis is indicated by broken lines. Based on ship observations during 1911-70.

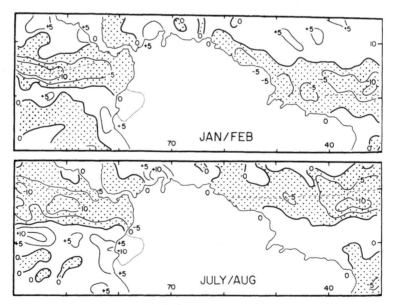

Figure 5. Surface divergence (10^{-6}s^{-1}) during January/February (top) and July/August (bottom). Convergent areas are shaded, and position of near-equatorial confluence axis is indicated by broken lines. Based on ship observations during 1911-70.

10

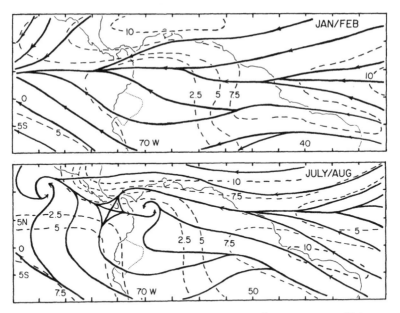

Figure 6. 'Gradient level' streamlines and isotachs (m s^{-1}) during January/February (top) and July/August (bottom); schematic after Atkinson & Sadler (1970).

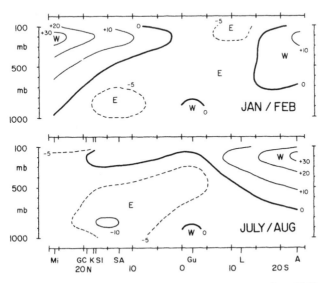

Figure 7. Meridional transects of zonal wind component along 80 W. a) January/February; b) July/August; Westerlies positive, isotachs in m s^{-1}. For approximate location of transect, see Figure 3.

11

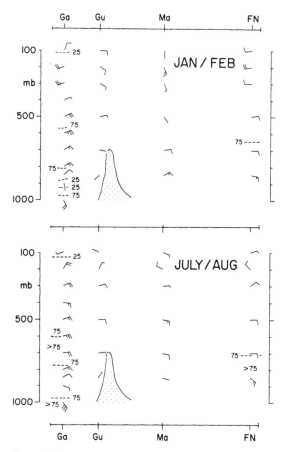

Figure 8. Zonal transects of upper-air resultant wind along the Equator. a) January/February; b) July/August. Short and long barbs denote 2.5 and 5.0 m s^{-1}, respectively. Isopleths are directional steadiness of wind, $\vec{V}/V$, in percent. For approximate location of transect see Figure 3.

(Figure 6). At higher levels, Easterlies with varying meridional components prevail over the Equatorial Americas throughout the year (Atkinson & Sadler 1970; Sadler 1975).

The seasonal shift of zonal wind regimes is illustrated in the meridional-vertical transects, Figure 7 (source: US Weather Bureau, ESSA, NOAA 1965-76). In January/February, the Northern hemispheric temperate-latitude Westerlies extend far equatorward in the high troposphere; the Southern hemisphere Westerlies are comparatively weak. In July/August, the Southern Westerlies are intensified, concomitant with the decay of upper-tropospheric Westerlies and strengthened lower and mid-tropospheric Easterlies in the

Northern hemisphere. During Northern summer, intensified Easterlies extend into the mid-troposphere in the equatorial belt, also South of the Equator. In the High Andes of Ecuador, strong Easterlies in the peak regions around the 500 mb level make this season unsuitable for climbing ventures, despite the generally clear weather. The Westerly wind component at 10-20 N reflects the development of an upper-tropospheric trough around the middle of the Northern summer (Hastenrath 1966, 1967). Lower-tropospheric Westerlies persist throughout the year near the Equator.

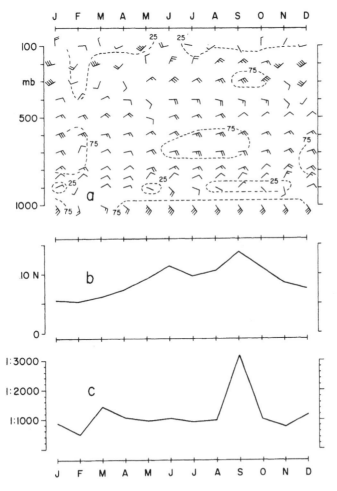

Figure 9. Galápagos. a) Annual variation of upper-air resultant wind. Symbols as in Figure 6; b) latitude position of surface wind discontinuity at 80-100 W; c) slope of wind discontinuity at 80-100 W. For approximate location of transect see Figure 3.

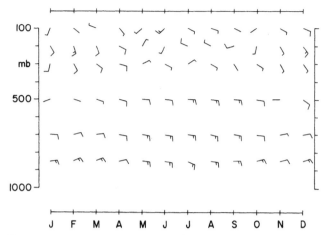

Figure 10. Manaus. Annual variation of upper-air resultant wind. Symbols as in Figures 6-7.

Zonal transects, Figure 8, illustrate upper-air wind conditions along the Equator (source: US Weather Bureau, ESSA, NOAA 1965-76; and unpublished data). The High Andes represent a powerful divide for the lower-tropospheric flow. To the East of the Andes, Easterlies prevail in the lower and mid-troposphere throughout the year, but reaching somewhat higher in July/August. More southerly directions develop in the surface flow during Northern summer. On the Pacific side of the Andes, Easterlies likewise dominate the middle and much of the lower troposphere. A very steady cross-equatorial flow is confined to the surface layer. Wind conditions in the upper troposphere over the South American continent during January/February are characterized by the northward directed outflow over the

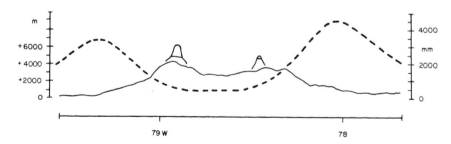

Figure 11. Annual precipitation totals (heavy broken lines) in a schematic zonal transect along approximately 1°S. Vertical exaggeration of topography is fivefold, as in Figure 2. For approximate location of transect see Figure 1.

14

greater Amazon basin, originating in the anticyclone over the Bolivian-Peruvian Altiplano; and by an associated col point in the area of Ecuador (Sadler 1975). During July/August, the equatorial belt is occupied by a buffer zone between Northern hemisphere Easterlies and Southern hemisphere Westerlies (Sadler 1975), manifesting itself in the transect by winds from northerly directions.

The annual march of upper-air resultant wind is illustrated in Figures 9 and 10 (source: US Weather Bureau, ESSA, NOAA 1965-76; and unpublished data) for two equatorial stations. At Galápagos, Figure 9, Easterlies prevail throughout the year in the middle and much of the lower troposphere. Upper-tropospheric features are consistent with Sadler's (1975) large-scale charts. More importantly, the short series of Galápagos upper-air wind measurements shows that the extremely steady, clockwise turning cross-equatorial surface flow has a typical thickness of only around 1 km. The poleward boundary of this current at the surface, and thus the slope of the interface between Southwesterlies below and Northeasterlies aloft (Figure 9), can be ascertained from long-term ship observations (Hastenrath 1977). It is noted that the strongly convergent cross-equatorial surface current over the Eastern Pacific (Figures 4 and 5) is extremely shallow.

Manaus, Figure 10, is representative of the Amazon basin, to the East of the Andes. Easterlies extend throughout most of the troposphere. Upper-tropospheric features are consistent with Sadler's (1975) large-scale charts. The broad Easterly flow in the lower troposphere, as represented by the 850 mb level, assumes a northerly component from October to April, but a switch to South of East is apparent in Northern summer.

In low latitudes, precipitation is the most important element in the regional climate. Reflecting the broad physiographic structure (Figure 2), the rainfall pattern in Ecuador is characterized by predominantly meridionally oriented isohyet configurations (Figure 11; source: Servicio Nacional de Meteorología e Hidrología 1972). A belt of maximum rainfall with annual totals of 2000 mm to more than 4000 mm is found at elevations around or under 1000 m. Precipitation decreases eastward towards the High Andes and the interior basins, which are characteristically dry, annual totals ranging from around 1000 mm to less than 500 mm. Precipitation increases eastward of the Cordillera Oriental, and rainfall is particularly abundant on the Amazonian slope of the Andes, where annual totals of 3000 mm to more than 5000 mm are reached at elevations around 1000-1500 m.

The annual march of rainfall in the Pacific lowlands in characterized by a simple annual variation with a distinct maximum around February/March, and a broad minimum from July to October. Major features in the annual rainfall variation are controlled by the large-scale vergence pattern in the lower-tropospheric flow, in that the distinct February/March rainfall

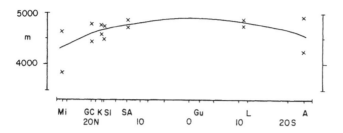

Figure 12. Annual mean elevation of 0°C isothermal surface, with annual range indicated by crosses. For location of profile and stations refer to Figure 3.

peak coincides with the southernmost seasonal migration of the convergence band over the neighboring East Pacific (Figure 5); and an extended dry season is centered around Northern hemisphere summer. Even the excessive rainfall during El Niño years (Hastenrath 1976a; Hastenrath & Heller 1977; Covey & Hastenrath 1978; Hastenrath 1978b) is seasonally anchored.

In most of the Sierra, a double variation of precipitation prevails, with the main and secondary maxima around March/April and October/November respectively. The main minimum occurs around July/August, and a secondary minimum around December/January. However, a distinct dry season as at the Pacific coast is not developed. In the Oriente, rainfall is abundant throughout the year, with largest values generally from April to November (Servicio Nacional de Meteorología e Hidrología, Ecuador, unpublished data).

The existence of a belt of maximum rainfall at intermediate, with a decrease towards higher elevations, such as depicted in Figure 11, is common in low latitudes (Hastenrath 1967). The seasonal precipitation regimes of the orographically complex Sierra and the Oriente are not well understood at present. Much of the precipitation-bearing weather systems in the highlands seems to propagate from the Amazonian side of the Andes; the aforementioned cross-equatorial surface flow and the nearby oceanic moisture source are conceivably important for the Pacific region.

The thermal pattern throughout the tropics is horizontally rather uniform. Temperature conditions in the free atmosphere over Ecuador (Figure 12; source: US Weather Bureau, ESSA, NOAA 1965-76) are characterized by an annual mean elevation of the 0°C isothermal surface somewhat below 4900 m, with an annual range of the order of 100 m, and a drop towards higher latitudes of either hemisphere. A heating effect related to the elevated mountain masses can be conjectured, but not substantiated from observations.

With the Ecuadorian Andes stretching broadly from 5°S to 1°N, an

16

asymmetry in insolation between northward and southward facing slopes results for mountains at the more southerly locations. While insolation would as a rule be enhanced on the equatorward slope, West-East asymmetries are less uniform. Zonal contrasts in radiation can be brought about by a pronounced diurnal variation in cloudiness, such that for example solar irradiance on westward facing slopes is reduced by increased afternoon cloud cover. This effect is conspicuous in some tropical regions with moderate cloudiness, but may be of subordinate importance for the perenially wet environment of the Ecuadorian Andes.

4 WESTERN CORDILLERA

C'est ainsi qu'au bord de la mer du Sud, après les
longues pluies de l'hiver, lorsque la transparance de
l'air a augmentée subitement, on voit paroître le Chimborazo comme un nuage à l'horizon: il se détache des
cimes voisines; il s'élève sur toute la chaîne des Andes,
comme ce dôme majestueux, ouvrage du génie de
Michel-Ange, sur les monumens antiques qui environ-
nent le Capitole.

Alexander von Humboldt, *Vues des Cordilleres,* 1810.

. . . when we got on to the dome, the snow was
reasonably firm, and we arrived upon the summit of
Chimborazo standing upright like men, instead of
grovelling, as we had been doing for the previous five
hours, like beasts of the field.

Edward Whymper, *Travels amongst the Great Andes
of the Equator,* 1892.

Glaciation of the mountains in the Western Cordillera is discussed in the following proceeding from South to North. For an overall orientation, reference
is made to Figure 1, which locates the 1:100,000 maps for the individual
mountain massifs. Fossil evidence of former glaciations is reviewed jointly
with the present ice extent, because glacier systems are viewed as temporally
continuous entities; and the chronology of glacial events is not known a
priori. The approach integrates field observation, air photo interpretation,
and evaluation of historical sources.

18

Photo 1. East side of Chimborazo at 4100 m. Vegetated triple moraine (III) complex belonging to predecessor of Chuquipoquio and Moreno glaciers (Dec. 1974).

Photo 2. East side of Chimborazo at 4050 m. Same moraine complex as in Photo 1 but further down valley (Dec. 1974).

Photo 3. Cut across moraine (III) profile, Southeast side of Chimborazo, at 3800 m. Same as shown in Photo 2. 1. Weathered tuff, about 4 m; 2. stratified debris, typical diameter of cm, possibly solifluctional, 30-50 cm; 3. rock debris, chaotic, 150 cm; 4. layers of pumice and sandy and clay-like tuff, 10 cm; 5. coarse material (Dec. 1974).

Photo 4. Carihuairazo from Northwest at 3350 m. Vegetated moraines (III) reach down to around 3500 m (Dec. 1974).

Photo 5. West side of Carihuairazo, view towards the West, at 4680 m. Bare moraine (II) (Dec. 1974).

Photo 6. West side of Carihuairazo, 4600 m. Glacier striations. Position somewhat above Photo 5, looking uphill. Box as scale has sides of 7 cm (Dec. 1974).

Photo 7. Chimborazo from Northwest at 3800 m. To the right Stübel (20) glacier, in middle Reiss (22), to left Spruce (1) glacier. Below Reiss glacier upper part of regenerated glacier (21) (Dec. 1974).

Photo 8. Northwest side of Chimborazo at about 5700 m. View down at regenerated glacier (21) between Reiss (22; right) and Stübel (20; left) glaciers (May 1975).

Photo 9. Chimborazo from Northeast from slope of Carihuairazo at 4550 m. To the right Abraspungo glacier (2), to the left Hans Meyer glacier (3) (Dec. 1974).

Photo 10. Chimborazo from Northeast, from slope of Carihuairazo at 4550 m; position as for Photo 9. To the right Abraspungo glacier (2), in middle Hans Meyer glacier (3), to left Reschreiter (4) and Wolf (5) glaciers (Dec. 1974).

Photo 11. Chimborazo from South at 3800 m. To the right location of Boussingault glacier (9) and ice lobes (10-12); the Southeast (13) and the remnant of the South-southeast (14) glaciers appear to the right and near the middle dome, respectively. The Little (15) and Great (16) South glaciers are visible below the main peak; and the Totorillas glacier (17) is to the extreme left (Dec. 1974).

Photo 12. Chimborazo from South at 3800 m, with position similar to Photo 11. To the right Southeast (13), in middle remnants of Southsoutheast (14) glacier, and to extreme left margin of Little South glacier (15) (Dec. 1974).

Photo 13. Chimborazo from Southwest at 4000 m. In the middle Little (15) and Great (16) South glaciers, to left Totorillas (17) glacier (Dec. 1974).

Photo 14. Chimborazo from Southwest at 4000 m. To the right Little (15) and Great (16) South glaciers, in middle Totorillas glacier (17) (Dec. 1974).

Photo 15. Chimborazo from Southwest at 4200 m. To the right Totorillas (17) glacier, in the middle valley leading to Debris glacier (18). Towards the left Thielmann glacier (19) (Dec. 1974).

Photo 16. Chimborazo from West at 4100 m. To right Thielmann (19), in middle Stübel (20) glacier (Dec. 1974).

Photo 17. Chimborazo from Northwest, at 4100 m. To extreme right margin of Thielmann (19), in the middle Stübel (20), and to left Reiss (22) glacier (Dec. 1974).

Photo 18. Carihuairazo from South at 4380 m. Rim of glacier tongues 4-6 of Map 1 is visible (Dec. 1974).

Photo 19. West side of Carihuairazo, 4650 m. Ice rim and glacial lake. Banding is apparent on ice cliff in background. Somewhat above Photo 6 (Dec. 1974).

Photo 20. Vegetated moraine (III) on Southeast side of Iliniza at 3700 m (Dec. 1974).

Photo 21. Northwest side of Atacazo, 3700 m. Profile through moraine of complex III: on top thick layer of weathered volcanic ash (Cangagua); beneath moraine core proper, with chaotic arrangement of rock debris (Dec. 1974).

4.1 CHIMBORAZO-CARIHUAIRAZO

Chimborazo (Figure 1 and Map 1), the highest, ice-clad peak of the Ecuadorian Andes has long fascinated naturalists and mountaineers alike. Large glaciers flow down from the gigantic ice cap in all directions. Earlier glaciations have left their traces far down the mountain slopes. Carihuairazo, itself ice-capped and peaking at more than 5000 m, is dwarfed by its gigantic neighbor. Yet, during earlier glaciations their ice streams joined, so that the two mountains are appropriately treated as one glacial entity.

Chimborazo and Carihuairazo were chosen as the first area of study in the December 1974 field work. Areas visited include the saddle between the two mountains; the Southwest side of Carihuairazo; and the Northwest flank, the lower portions of the South side, and the Eastern sector of Chimborazo. In May 1975, the summit of Chimborazo was climbed on the conventional route from the Northwest.

Lowest moraines (III) have a height of the order of tens of metres, they are covered by a mantle of volcanic ash several metres thick, and carry vegetation. These moraines extend from around 4600 m down to lowest elevations of less than 3600 m in the Southeast sector of Chimborazo, contrasting with values of only around 4200 m to the Northwest. In context it is noted that the base of the mountain is higher in this sector. A similar asymmetry is indicated for Carihuairazo. This moraine complex manifests that ice streams from Chimborazo and Carihuairazo combined in the region of the intermediate Abraspungo saddle. From here, ice originating on both peaks flowed down as one stream towards the Northeast into the Mocha Valley. Likewise, ice from both mountains was incorporated into one large ice stream directed from the saddle area towards the Northwest.

Moraines of this complex are illustrated in Photos 1-4 for selected portions of the mountains. In the Southeast sector of Chimborazo a large group of vegetated moraines with three pairs of parallel ridges extends from below the present glaciers 7 and 8 down to around 3800 m. Photo 1 gives an impression of this huge moraine complex at 4050 m, and Photo 2 is taken at a somewhat lower site, with view to the lower end of the moraine complex. A cross-section through the front portion of the lowermost moraine is shown in Photo 3. Grass stocks on a thick cover of deeply weathered volcanic ash (layer 1). The layered debris material below (layer 2) is suggestive of solifluction characteristic of a periglacial environment. The entity of very coarse and chaotically arranged blocks (layer 3) is the moraine core proper. Interbedded thin layers of pumice and tuff indicate volcanic activity at some time in the course of the moraine deposition. Photo 4 shows a moraine pair ascribed to this complex (III) on the Northwest side of Carihuairazo extending down to about 3500 m.

Moraines (II) of distinctly different appearance are found at elevations of 4700-4400 m: the moraine material proper is bare, without ash mantle and vegetation. The height of the moraine ridges is of the order of tens of metres. Specimens of this complex (II) were observed in the terrain on the Southwest flank of Carihuairazo, and on the Northwest side and the Eastern sector of Chimborazo. Pertinent inference for other parts of the mountain massif was possible from air photography. Photo 5 shows a bare moraine ridge ascribed to complex II, on the West side of Carihuairazo below 4600 m. This area is below the present ice lobes 7, 8, 9 (Map 1). Glacier striations on bare rock somewhat above the location illustrated in Photo 5 are shown in Photo 6.

Another complex of moraines (I) of fresh appearance without ash and vegetation cover is found only tens of metres below the present ice rim, at elevations of 4800-4700 m. Moraines of this complex (I) were observed in the terrain in the area visited. Inference for other portions of the mountains was possible from air photographs. Historical photographs for the Chimborazo-Carihuairazo region are not informative on the age of these moraines. From photographic documentation at Antisana (section 5.13) and the similarity

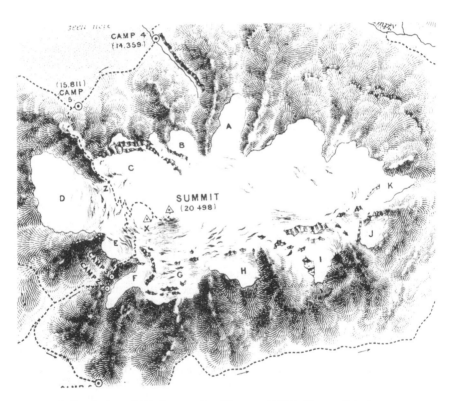

Figure 13. The glaciers of Chimborazo after Whymper (1892). Names of glaciers: A – Abraspungo; B – Spruce; C – Reiss; D – Stübel; E – Thielmann; F – Debris; G – Tortorillas; H – Humboldt; I – Boussingault; J – Chuquipoquio; K – Moreno.

in appearance and altitudinal arrangement it seems that the ice rim may have been close to moraine complex (II) during the 19th, and that complex (I) formed during the 20th century.

Consistent with the fossil moraine morphology, the recent glaciation of Chimborazo shows a marked asymmetry in that glaciers extend to lowest elevations of around 4700 m in the Southeast and East sectors, whereas the ice rim stays as high as 5000 m and more to the Northwest.

Little definitive historical information on recent glacier variation can be derived from the reports of travellers during the past centuries. Wagner (1870) compared Humboldts's (1802) and Boussingault's (1831) observations of the snowline on Chimborazo (Appendix II: General 13), but uncertainties in height determinations and the sector referred to preclude interpretation in either relative or absolute terms. Reiss & Stübel from their 1870-74 expedition (Reiss & Stübel 1892-98; Dietzel 1921) gave an account of the snowline at Chimborazo in considerable detail (Appendix II: General 16; Chimborazo-Carihuairazo 3, 4). They seem to have been the first ones to recognise an asymmetry, with lowest values in the East and a highest ice rim to the Northwest. Their figures may indicate an ice retreat during the last 100 years,

Table 1. Nomenclature of glaciers on Chimborazo.

Whymper	Meyer	Sauer	Map 1
Spruce	Spruce	Spruce	1
Reiss	Reiss	Reiss	22
–	–	Lea Hearn	21
Stübel	Stübel	Stübel	20
Thielmann	Thielmann	Thielmann	19
Debris	Trümmer	Trümmer (escombros)	18
Tortorillas	Totorillas	Totorillas	17
Humboldt	⌈ Grosser Süd	⎫ Walter Sauer?	16
	∣ Kleiner Süd	⎬	15
	∣ Südsüdost ⌉ or	⎫ Humboldt	14
	⌊ Südost ⌋ Humboldt	⎭	13
–	–	Carlos Pinto	12
		sketched, but unnamed	11
		Nicolas Martínez	10
Boussingault	Ostsudost or Boussingault	Boussingault	9
Chuquipoquio	Chuquipoquio	–	8
Moreno	Ost or Moreno	García Moreno	7
two sketched, but not named	⌈ Wolf	Theodoro Wolf	6
	∣ –	Carlos Zambrano	5
	∣ Reschreiter	Reschreiter	4
	⌊ Hans Meyer	Hans Meyer	3
Abraspungo	Abraspungo	Abraspungo	2

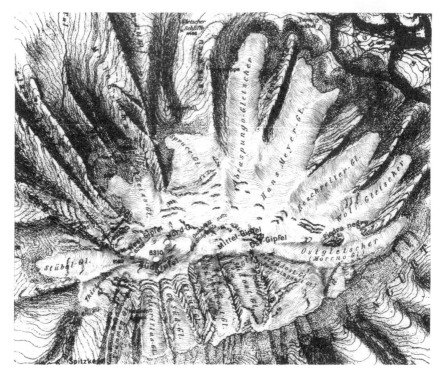

Figure 14. The Glaciers of Chimborazo after Meyer (1907).

were it not for the uncertainty in height determinations. Whymper's (1892)
report on snow conditions in 1880 (Appendix II: General 18, 19) is in context not useful.

  ⹁ Whymper (1892) published the first sketch of glaciers on Chimborazo in
1880, his map being reproduced as Figure 13. He gave a more detailed sketch
of the Debris Glacier (18). Chimborazo was the prime target of investigation
for Meyer's (1907) expedition in 1903. His map pertaining to an inventory
of recent glaciers on this mountain is shown as Figure 14. Sauer (1965,
1971) published a further sketch of the Chimborazo glaciers (Figure 15).
The nomenclature is compared in Table 1.

  Chronically clouded conditions, logistic difficulties for certain sectors of
the massif, and poor definition of ice lobes, rather than prominent secular
changes, may be major factors in the discrepancies borne out by Table 1.
There is almost complete agreement of nomenclature for the Northern and
Southwestern sectors of the mountain. Sauer's 'Carlos Zambrano-Glacier'
(5) can from available air and ground photographs be regarded as a sub-

22

Figure 15. The glaciers of Chimborazo after Sauer (1971). Names of glaciers: GM – García Moreno; TW – Theodoro Wolf; CZ – Carlos Zambrano; Rr – Reschreiter; HM – Hans Meyer; Abr – Abraspungo; Sp – Spruce; Re – Reiss; LH – Lea Hearn; St – Stuebel; The – Thielmann; Trü – Trümmergletscher; To – Totorillas; WS – Walther Sauer; Hu – Humboldt; CP – Carlos Pinto; NM – Nicolás Martínez; Bou – Boussingault; N, Oe, S, M, E: North, West, South, Middle, and East peaks; P – Piedra Negra; C – Catedral; MAM – Murallas Augusto Martinez.

entity of the Wolf (6) Glacier. While I visited adjacent sectors, this area was not in line of sight.

Inconsistencies in nomenclature for the South and East sides of the mountain are less satisfactorily reconciled. Whymper maps one large 'Humboldt Glacier' on the South flank of the massif, where Meyer distinguishes four separate ice streams (13-16); and Whymper's map lacks detail in the Northeast sector. It is not believed that this reflects secular change. More remarkable are the differences between Meyer's and Sauer's maps. The correlation attempted in Table 1 is doubtful in some areas. Field observations during December 1974, supported by air photographs, are in part consistent with Meyer's and Sauer's sketches. Sauer's nomenclature and sketch for the sector between the Moreno (7) and Totorillas (17) Glaciers may largely refer to details of the ice rim. Separate lobes possibly corresponding to the ones referred to by Sauer are apparent both in the field and on air photographs. With reference to Meyer's sketch this may reflect secular disintegration of the ice cover. A 'Chuquipoquio' Glacier (8) is lacking on Sauer's sketch. Remnants of this small glacier may only represent a perennial ice field. Observations from a distance in December 1974 do not resolve this question. The Southsoutheast glacier (14) of Meyer has lost definition. And the so-called 'Little' South glacier (15) now appears more powerful than the 'Great' South glacier (16). At variance with Whymper and Meyer, Sauer named a 'Lea Hearn Glacier' in the area between the Stübel (20) and Reiss (22) Glaciers. An ice tongue (21) presumably referred to by this name was observed in the area during the 1974-75 field work (Photo 8). This glacier seems to regenerate at comparatively low elevations from ice masses supplied over steep cliffs above. Photos 7 through 17 show overlapping aspects for all sectors of Chimborazo except the Northeast. In conjunction with the clockwise numbered glaciers/ice lobes in Map 1, these allow an orientation on the present glaciation.

In qualitative terms, a continued ice retreat can be ascertained for the Northwest sector of Chimborazo, based on Meyer's (1907) photographs in 1903, the unpublished photographs of Nicolás Martinez around 1906 (courtesy of 'Nuevos Horizontes'), Eichler's (1952) pictures of around 1950, and the 1974-75 field observations. Approaches to other sectors of the mountain have remained less popular — a fact related to the more abundant precipitation, chronically poor weather, and the more vigorous glaciation.

For Carihuairazo, hisotrical photographs did not permit a reconstruction of long-term glacier behavior. The aforementioned moraines of fresh appearance of complex II, in context suggest an ice retreat since sometime in the past century. Photographs of Carihuairazo from the Northwest taken by Nicolás Martinez in 1906 (courtesy of 'Nuevos Horizontes') show fields of snow or ice below steep cliffs, similar to the present appearance. However, secular variations cannot be safely ascertained. Three ice tongues (1, 2, 3) can be made out above cirques in the eastern sector. To some mountaineers, the tongue (2) is known as 'Glaciar de Carbonería', and tongue (3) as 'Glaciar de Salasaca'. Three small glacier tongues (4, 5, 6) can be identified to the Southeast above a valley tributary to Rio Mocha (Photo 18). These were tentatively named Hoinkes, Kinzl, and Troll. Three rather rudimentary ice lobes (7, 8, 9) were found on the West side. A photograph by P. Grosser in 1902 (source: Meyer 1907: Vol. 2, Plate 17A) shows a considerably larger ice extent in this sector, at the beginning of the century. The ice rim and a glacial lake in this general area is shown in Photo 19. Some perennial ice fields seem to persist below cliffs on the North side of Carihuairazo.

4.2 ILINIZA

The twin peaks of Iliniza (Figure 1 and Map 2) loom to well above 5000 m, and the Southern one is presently strongly glaciated. A rich geomorphic legacy attests to a formerly much larger ice extent. In December 1974 the peak region was visited by the now conventional route from the Northeast. This included a reconnaissance of the saddle between the two peaks and the adjacent portions of the East and West flanks of the Southern (main) dome. On two subsequent trips the massif was approached from the Southeast and West sides. Inclement weather and clouds hampered the field observations.

Moraines (III) with a height of the order of tens of metres reach down to elevations below 3600 m, with a largest extent apparently in the Southeast quadrant and somewhat higher limits to the West. All of these low-reaching moraine ridges are covered by several metres thick layers of volcanic ash and abundant vegetation. Different phases are indicated within this moraine complex, for example, on the Southeast and North sides of the massif. A moraine of this complex at 3700 m on the Southeast side of Iliniza is shown in Photo 20.

A distinctly different set of moraines (II) of somewhat smaller height, is found upward of about 4300 m, and on the South side above 4200 m. This complex is characterized by bare moraine material, that is the absence of ash mantle and vegetation. A moraine of this group was studied in situ on the Northeast side, but corollaries on the South and West sides could be verified from the distance and by means of air photographs. The upper ends of this moraine complex seem to lie around 4600-4700 m.

Still higher up, and only some tens of metres below the present ice rim around 4800 m, yet another complex of moraines (I) was encountered, again bare, and rather fresh in appearance. Complementing the field survey on the North side of the main dome, corollaries of this moraine complex could be detected from air photography on the Southeast side.

Moraine ridges of the various complexes (I-III) are broadly aligned in a topographically consistent way with the recent glaciation. Moraines of complexes III and II in particular originate at the saddle, with the corresponding glaciers apparently having been fed from both peaks jointly.

The modern glaciation is confined to the somewhat higher Southern peak. Five distinct tongues (1, 2, 3, 4, 5) are apparent to the East and Southeast. And further three tongues (6, 7, 8) are indicated on the Southwest to West flanks. Two glacier tongues (9, 10) extend from the South peak towards the saddle. Ice extends to below 4800 m to the Southeast, but stays at higher elevation on the West side. Ice seems to be least extensive to the West. A preference of the Southeastern over the Western sector is indicated for both the earlier and the present glaciations. The present ice rim is separated from the innermost moraines.

Reference to the modern glaciation is contained in expedition reports from the past century. Wagner's (1870) entry for the lower limit of perennial snow on Iliniza in 1858-59 (Appendix II: General 12) cannot be reliably evaluated in terms of elevation. Reiss and Stübel (Stübel 1897; Reiss & Stübel 1892-97) visited the mountain during their 1870-74 expedition (Appendix II: General 16; Iliniza 3, 4). They report glaciers descending from the North side of the South dome towards the saddle between the twin peaks; more importantly a large glacier is described as extending from the saddle itself towards the West, down to 4484 m, with a snow line at 4658 m. Their figure of about 4800 m for the saddle compares reasonably with the modern topographic sheet, and therefore provides a useful reference for their other elevation estimates. Their observations thus pinpoint a much larger ice extent than at present. In fact, their elevation value for the glacier snout corresponds to bare moraines of fresh appearance in the aforementioned area, apparent both from field observations and air photographs. It is concluded that moraines of complex II as described above were nearly in contact with the ice rim around 1870.

Whymper's (1892) narrative from his visit in 1880 (Appendix II: Iliniza 5) cannot be appreciated quantitatively. Meyer (1907) on his 1903 expedition did not visit the mountain and only observed the peaks from afar (Appendix II: Iliniza 6). His account of small ice tongues on the North peak does not allow quantitative comparison. Photographs from more recent decades obtained from the mountaineering community in Quito are helpful towards a general glacier inventory, but do not detail secular variations.

4.3 CORAZON

The Corazón (Figure 1 and Map 3) is several hundred metres lower than the neighboring Ilinizas to the South, and it thus does not take part in the present glaciation. Similar to other mountains of volcanic origin in the Western and Eastern Cordilleras, the remnants of the caldera opens towards the West. Glacial morphology was observed from a distance on the readily accessible South and East sides, and during December 1974 moraines were studied in situ in the Northern sector.

Moraines (III) tens of metres high reach down to below 3800 m, and in the Northwest and West to about 3200 m. These carry a thick mantle of volcanic ash and abundant vegetation. Different phases of moraines (III) are indicated in the sectors from Northwest to East, and South.

While the peak region was not visited, a group of seemingly bare moraines (II) was identified from air photographs, to the East and South extending down to about 4400 m and on the West side of the crest possibly to 4200 m and less. The lack of an ash mantle and the similar elevation suggests a correlation with moraine complex II on the neighboring Ilinizas. Moraines corresponding to complex I at the Ilinizas could not be detected; in fact, even the peak of Corazón stays below the present glacier limit on the Ilinizas. In contrast to the Ilinizas. glaciation seems to be favored on the Western side of the crest. This feature is indicated for other volcanic peaks with westward opening caldera remnants, and may reflect a topographic control.

Corazón is one of several presently not glaciated mountains in the vicinity of the capital city of Quito, which are of easy access and have been mentioned repeatedly by visitors in the course of the past more than two centuries. Mountains such as Corazón, Atacazo, Pichincha, Rumiñahui and Pasochoa, are of particular interest in that they are close together, their peaks cluster over a relatively narrow elevation range, and most importantly, they are only little lower than nearby peaks presently glaciated. Whereas elevation estimates of snowline and glacier snouts from earlier centuries are as a rule dubious, the presence or absence of perennial snow and ice is definitive information.

The municipal records of Quito (González Rumazo 1934a) contain descriptions which can be interpreted to the effect that Pichincha, Corazón, Ilinizas and possibly even Atacazo carried perennial snow around 1535. However, the archives are inconclusive in the details, because of the lack of mountain and place names (Appendix II: General 1).

The French Academician La Condamine (1951) gives a vivid description for a time as early as 1738 (Appendix II: General 5; Corazón 3). La Condamine is unmistakably aware of the difference between merely temporary and perennial snowlines; for the latter his account would correspond to about 4700 m. Around the same time, Velasco (1841-44) describes the mountain as having perennial snow (Appendix II: General 6; Corazón 5).

Humboldt (1810, 1853) describes and depicts the mountain as capped by perennial snow in 1802 (Appendix II: General 8, 9; Corazón 7). For around 1858 (Appendix II: General 11, 13) both Villavicencio (1858) and Wagner (1870) list Corazón among the mountains with perennial snow. Orton (1870) mentions Corazón, as well as Pichiñcha and Rumiñahui without commenting on their snow/ice conditions, whereas Cotacachi is described as perpetually snow-clad. 22 summits are said to be covered with perennial snow, but not identified (Appendix II: General 14). In context Orton's account suggests that snow conditions on these mountains may have been marginal, but this is speculative. Reiss' (Dietzel 1921) and Reiss & Stübel's (1892-98) accounts for 1870-74 (Appendix II: General 15, 16; Corazón 12) suggest perennial snow on Corazón. However, the statements are rather weaker than La Condamine's for 1738, in particular when viewed in perspective with Reiss & Stübel's (1892-1898) description for nearby Iliniza (Appendix II: Iliniza 4) in about the same year.

Whymper's (1892) descriptions for 1880 (Appendix II: General 18; Corazón 15) could be interpreted as indicating less snow than at the time of Reiss' (Dietzel 1921) visit a decade earlier (Appendix II: General 15), although Whymper's formulation leaves doubt. Wolf (1892) still lists Corazón among the mountains with perennial snow (Appendix II: General 21, 22; Corazón 18). Meyer (1907) in 1903 (Appendix II: General 23) considers the perennial snow cover on Corazón as marginal. And perennial snow seems to have disappeared gradually in the decades since then.

4.4 ATACAZO

The peak of Atacazo (Figure 1 and Map 4) stays about another 200 m below that of the neighboring Corazón, with a corresponding further impoverishment of fossil glacial morphology. Similar to Corazón and other volcanic peaks, the remnant of the caldera opens towards the West. In December 1974, field trips were undertaken to four separate portions of the mountain, two of the travels extending to the crest. Again, obstruction by cloud posed a handicap.

Moraines (III) with a height of tens of metres, an ash mantle of several metres and abundant vegetation extend down to below 3600-3400 m in the Southeast, and possibly somewhat lower in the Northwest sector. This would be in similarity to Corazón. Indications for separate moraine phases (III) appear in the Northwest sector. Photo 21 of a moraine profile on the Northwest side of Atacazo at 3700 m illustrates a deep layer of weathered volcanic ash overlying the rock debris of the moraine core proper.

Moraine ridges without ash mantle and vegetation cover — that might be correlated with complex II of Corazón and the Ilinizas adjacent to the South — could not be found in the terrain. Inference from air photography of such a feature on the Southern end of the Atacazo crest is dubious. This state of affairs seems consistent with the distinctly lower elevation of Atacazo.

It is doubtful, although not impossible (González Rumazo 1934a) that Atacazo may have carried perennial snow around 1535 (Appendix II: General 1). This prospect should be viewed in context with other reports in the municipal records of Quito (González Rumazo 1934a), which repeatedly mention a 'sierra nevada de pinta' in 1540 and 1550 (Appendix II: General 2, 4). 'Pinta' must be the town of Pintag some 25 km to the Southeast of Quito (see also the map in Schottelius 1935-36). The descriptions may refer to Sincholagua and the mountains to the East of Pintag, which rise to little over 4400 m but are presently not glaciated. Certainly, the term 'sierra nevada', indicates more extensive snow cover than at present in the Pintag region. The account of Gonzalo Pizarro's expedition to the Oriente in 1541 (González Suárez, 1890-1903; Bonifaz, 1971) also describes the abundant snowfall and cold weather encountered in the Eastern Cordillera (App. II:

General 3). If indeed the mountains to the East of Pintag are meant by 'sierra nevada', then the snow and ice conditions in the high Andes must have been comparatively severe. The existence of perennial snow cover, though not of glaciers, on Atacazo around 1530-50 seems consistent with such a scenario.

Humboldt (1874a) in 1802 found the mountain lacking perennial snow (Appendix II: General 8). Villavicencio's (1858) allegation of perennial snow on Atacazo (Appendix II: General 11) in context appears dubious. Karsten (1886) for the 1880's, Wolf (1892) for around 1890, and Meyer (1907) for 1903 report the mountain as free of snow (Appendix II: General 20, 23; Atacazo 9).

4.5 PICHINCHA

The highest peak of this extended mountain massif (Figure 1 and Map 5) is only about 100 m lower than that of Corazón to the South. As with Corazón, the mountain is presently not glaciated, although snow may persist for prolonged periods in sheltered locations. The peak region was visited on two occasions in January and in May 1975, both times by the conventional route from the East. Field trips during December 1974 covered the South and North sides of the mountain. Bad weather in the peak area often prevented observation even at short distance.

Moraines (III) with a height of the order of tens of metres and carrying an ash mantle of several metres and abundant vegetation extend down to 3500-3400 m in the Southern sector, and possibly to even lower elevations on the North side, although forms here are doubtful. A nesting of different moraine complexes (III) is apparent in the Southern sector. Moraine morphology in the precipitous and densely wooded terrain in the Western quadrants is difficult to ascertain from air photography.

Features interpretable as moraines – without a volcanic ash cover – are apparent on air photographs of the peak region at elevations around 4600 m down to possibly 4200 m. On my ascents to the peaks I must have been in line of sight of these features, but they were shrouded by clouds. At least some of these ridges may correspond to the complex II of bare moraines on Corazón and the Ilinizas.

The municipal records of Quito (González Rumazo 1934a) suggest that Pichincha was perennially snow-capped around 1535 (Appendix II: General 1). A definitive report of perennial snow cover on Pichincha (Rodriguez de Aguayo 1965) exists for around 1570-74 (Appendix II: Pichincha 1).

Pertinent observations about snow and ice conditions around 1740 (Appendix II: General 5; Pichincha 4, 5, 6, 7, 8) are provided by the expedition of the French Academicians (Juan & Ulloa 1748; Bouguer 1749; La Condamine 1751), and Velasco's (1841-44) account. The mountain evidently reached into the region of perennial snow.

Humboldt (1853, 1874a) attests to this state of affairs still in 1802 (Appendix II: Pinchincha 9, 10), although he confuses the two peaks Rucu-Pichincha and Guagua-Pichincha, as is apparent from his picture (Appendix II: Pichincha 11). For around the middle of the 19th century, Villavicencio (1858) reports Guagua-Pichincha as free of snow, but still lists Pichincha among the mountains with perennial snow (Appendix II: General 11; Pichincha 12). For around 1858 Wagner (1870) also cites Pichincha as carrying perennial snow (Appendix II: General 12, 13). Jameson (1861) in 1859 describes Pichincha as barely reaching the snow line (Appendix II: Pichincha 16). By the time of Reiss & Stübel's expedition (Reiss & Stübel 1892-98; Dietzel 1921) in 1870-74, and Dressel's (1877) travels in the 1870's snow conditions seem to have become marginal (Appendix II: General 15, 16, 17). A similar state of affairs is indicated for the visits of Whymper (1892) in 1880 and of Meyer (1907) in 1903 (Appendix II: General 18; Pichincha 21). A development towards less persistent snow seems to have continued since the turn of the century.

Figure 16. View of the Southeast side of Cotacachi peak in 1880 (from Whymper, 1892).

4.6 COTACACHI

This mountain (Figure 1 and Map 6) in the Western Cordillera still carries perennial ice. Air photographs and planimetric maps exist for the area, and a topographic map of older vintage at scale 1:25,000 is available for the Southern part of the mountain (Appendix I). Inference is based on air photography and consultation with local mountaineers.

Large, seemingly vegetated moraines (III) at lower elevations are apparent from air photographs. These may correspond to the complex (III) of ash-covered and vegetated moraines on other mountains.

At somewhat higher elevations, another set of moraine ridges (II) can be distinguished on air photographs, this one apparently bare, that is without ash cover and vegetation. A correspondence with the complex of bare moraines (II) on other mountains is suggested.

Small moraine arcs (I) of a fresh appearance can be made out from air photography in immediate vicinity of the present ice rim. This is again in similarity with other glaciated peaks.

Two small glaciers are reported to exist presently, extending to the West and South, respectively. The glacier tongue shown on Whymper's (1892) etching (present Figure 16) as viewed from the Ibarra sector, is no longer to be seen. General observations on the modern glaciation of Cotacachi have been made by Reiss & Stübel (1892-98) and Stübel (1897) in 1870-74 and by Whymper (1892) in 1880 (Appendix II: General 16; Cotacachi 2, 3). A picture in Wolf (1892:98) shows little detail. However, the etching of the East side of the peak in Whymper's book, reproduced as Figure 16, suggests a much larger ice extent in the latter part of the past century than presently. The fresh moraine arcs below the present ice rim may be related to those conditions. Stübel's (1897) account (Appendix II: Cotacachi 2) would also warrant quantitative comparison with present conditions.

4.7 CHILES

This volcano straddles the border of Ecuador and Colombia. A topographic map of older vintage at scale 1:25,000 is available for the Ecuadorian part, and air photography flown in 1978 for all of the mountain (Appendix I). This does not allow satisfactory assessment of moraine morphology, so that no map is presented here. The air photographs and a report from recent visits indicate that the mountain is presently free of ice, although perennial ice apparently existed earlier (Appendix II: General 20, 21).

5 EASTERN CORDILLERA

... estamos del real del señor gobernador Gonzalo
Pizarro, ducientas leguas o más por la tierra, todas sin
camino ni poblado, antes muy bravas montañas, las
cuales hemos visto por experiencia e vista de ojos
veniendo por el agua abajo en el dicho barco y canoas
padeciendo grandes trabajos y hambre.

Fray Gaspar de Carvajal, *Petition to Francisco de
Orellana,* 4 January 1542 (Archivo de Indias).

Patterned after the presentation in Chapter 4, the mountains of the Eastern
Cordillera are reviewed proceeding from the Southern hemisphere towards
the Equator. Figure 1 locates the 1:100,000 maps for the individual mountain massifs. This ensemble includes some peaks situated in the Eastern part
of the Interandean Depression, rather than the Eastern Cordillera proper.
Again an integrated approach is taken towards fossil and recent glaciation,
combining field work, air photography and literature research.

5.1 CERRO RASULLANA

The region to the Southeast of Cañar (Figure 1 and Map 7) is of moderate
elevation and does not take part in the present glaciation. A geological mapping of the area has been completed by the Dirección General de Geología
y Minas (1975), and Map 11 is not meant to add to this survey. Dr R. Bristow kindly placed his then unpublished map sheets and air photographs at
my disposal. A radiocarbon date is used in the absolute dating of stratigraphy.
This is of particular interest, inasmuch as it so far provides the only bracket
on the age of moraines. The highest part of the area was visited in June 1975.

Moraines (III) covered by volcanic ash and abundant vegetation extend
down to below 4000 m, and at one place to almost 3800 m. There is some

indication of preference for Southern slopes. Some nesting of moraines (III) is also apparent. These moraines are considered as possible corollaries of moraine complex III in other mountain regions, such as Atacazo. The rather higher inferior limit of these moraines as compared to Atacazo – a mountain of comparable elevation – is understood in the context of lower peak elevation and perhaps a topography less effective as ice reservoir in the Cerro Rasullana region.

Bare moraine ridges – without ash mantle and vegetation – that could be correlated with moraine complex II on other mountains, could not be detected. This is found consistent with the absence of such features on Atacazo, a mountain with somewhat higher elevation.

The aforementioned moraines in the Cerro Rasullana area overlie the pyroclastics of the Tarqui formation. Radiocarbon dates of 24,900 and 34,300 B.P. have been obtained from samples at the base of the Tarqui formation in the neighboring Azogues area (Bristow 1973; Dirección General de Geología y Minas 1974, 1975). By inference, this would bracket the moraines of complexes III, II and I, as all being younger than 25,000 B.P. However, an appreciable tolerance may have to be allowed for the varying age of Tarqui deposits at different locations. This in turn limits the validity of the 25,000 B.P. date as a maximum age of moraine complex III.

5.2 SOROCHE AND COLAY

These two mountains lying in a remote Southern region of the Eastern Cordillera are rarely visited. Topographic maps and a geological mapping are available, as well as air photographs of both 1956 and 1977 (Appendix I). Soroche, also called Ayapungo, peaks around 4730 m; and Colay, also called Achipungo, at 4630 m.

No map is presented here, because the air photography does not warrant assessment of moraine morphology and recent ice extent without verification in the field. There also seems to be room for improvement of the geological map. Reportedly, these mountains carry small glaciers. The air photographs show much smaller white patches in 1977 than in 1956. While this is suggestive of secular ice recession, the contribution of short-term variations in snow cover cannot be singled out.

5.3 SANGAY

This active volcano (Figure 1) on the wet Amazonian side of the Andes carries perennial snow, although bare ash makes up the peak region. Some reference to snow and ice conditions is found in the accounts of Velasco

(1841-44) for 1727-67, of Spruce (1861) for 1857, of Reiss & Stübel (1892-98) for 1870-74, of Whymper (1892) for 1880, and of Moore (1930) for 1929 (Appendix II: General 16; Sangay 1, 2, 4, 5). Volcanic activity can be expected to be a factor for the ice conditions. In fact, the mountain was found free of snow in 1976 (Marco Cruz, Quito, personal communication 1977). Few mountaineering parties have visited the region, as logistics are cumbersome. There are no air photographs and topographic maps of the area; accordingly, it was not included in the 1974-75 field survey. The far eastward location would make the pattern of recent and earlier glaciations interesting in comparison with other parts of the Ecuadorian Andes.

5.4 CUBILLIN

This mountain with elevations around 4500 m to the South of El Altar is rarely visited. Reportedly it carries perennial ice. Air photographs are available of both 1963 and 1977, but no topographic map (Appendix I). Since moraine morphology and recent ice extent could not be adequately determined from the air photographs, no map is presented here. As for Soroche & Colay (Section 5.2), the air photographs suggest an ice recession, barring short-term variations in snow cover.

5.5 EL ALTAR

El Altar, or Capac Urcu, is the most beautiful mountain of the Ecuadorian Andes. As might be expected from the peak elevations well above 5000 m and the location in the moist Eastern Cordillera, El Altar (Figure 1 and Map 8) is at present strongly glaciated, with a corresponding wealth in fossil glacial morphology. The mountain is of volcanic origin, with the remnant of the caldera opening towards the West, similar to several other mountains, such as Corazón, Atacazo, Rumiñahui and Pasochoa. A field trip to the caldera sector of El Altar materialized in May 1978. Air photographs are evaluated with reference to a planimetric sheet, contours being available only for a small portion of the Collanes Valley (Appendix I).

Apparently vegetated moraines (III) are found at lower elevations in various sectors of the massif. The jungles to the East may well hide features that are not readily recognized from air photographs. Extensive bare rock surfaces with embedded lakes suggestive of glacial erosion are particularly conspicuous to the South of the mountain. Platforms and precipices appear stratigraphically controlled.

At somewhat higher elevations, seemingly bare moraine ridges (II) can be made out from air photography, in particular in the Western and Southern

34

Photo 22. Upper Collanes Valley of El Altar, four lateral moraines (III), highest around 4050 m, lowest around 3800 m, on Southern slopes of valley (May 1978).

Photo 23. Collanes Valley of El Altar, down valley from Photo 22 around 3800 m. Moraine (III) with ash-covered and vegetated boulder.

Photo 24. The Caldera Glacier (2) of El Altar in 1872 according to a painting by Rafael Troya (source: Meyer 1907, Vol. 2). Note moraines (II) below rock face.

Photo 25. The Caldera Glacier (2) of El Altar in 1902, according to a photograph by Paul Grosser (source: Meyer 1907, Vol. 1).

Photo 26. The Caldera Glacier (2) of El Altar in 1903, according to a painting by R. Reschreiter (source: Meyer 1907, Vol. 2).

Photo 27. The Caldera Glacier (2) of El Altar in May 1978. Moraines (II) below rock threshold as in Photos 24-26.

Photo 28. The Caldera Glacier (2) of El Altar in May 1978. Telephoto from same position as Photo 27.

Photo 29. Northwest part of the caldera of El Altar at about 4350 m. Bare moraine (II) below glacier tongue (1) (May 1978).

Photo 30. View from atop moraine (I) at caldera exit in about 4150 m to Caldera Glacier (2) and lake (May 1978).

Photo 31. North side of Quilindaña at 4000 m. View down valley (northward) on vegetated moraines of complex (III) enclosing swampy terrain (Dec. 1974).

Photo 32 (left top). Cotopaxi viewed from Antisana in the Northwest (June 1975).
Photo 33 (left middle). North side of snout of glacier 12 at Antisana (Great West
glacier; Los Crespos) at 4500 m. In the foreground vegetated moraine (III), then bare
moraine ridge (II), behind that younger moraines (I) and present ice rim. Dotted line
indicates ice extent in 1903. In comparison with Photo 34 note drastic ice recession at
snout and side, and formation of younger moraines, since the beginning of the century
(June 1975).
Photo 34 (left bottom). North side of snout of glacier 12 (Great West glacier; Los
Crespos) at 4500 m. Same position as Photo 33, but taken by H.Meyer in 1903.
Photo 35 (top). North side of snout of glacier 12 (Great West glacier; Los Crespos) at
4500 m. Position similar to Photo 33, but telephoto. Dotted line indicates ice extent
in 1903 (June 1975).
Photo 36 (bottom). North side of snout of glacier 12 (Great West glacier; Los Crespos)
at 4800 m, with fresh moraine in foreground (June 1975).

Photo 37. Moraine (I) on North side of glacier 12 (Great West glacier; Los Crespos) at 4800 m. Dotted line indicates ice extent in 1903. By comparison with Photo 38, note formation of young moraine ridge and thinning of glacier since the turn of the century (June 1975).

Photo 38. North side of glacier 12 (Great West glacier; Los Crespos) at 4800 m. Position similar to Photo 37, but taken by H. Meyer in 1903.

sectors, but also to the North. Comparison of elevations is precluded by the lack of topographic maps. Moraines (I) close to the present ice rim are apparent from air photography, both in the caldera, and on the East side.

The westward opening caldera is the best accessible portion of the mountain, and has been more commonly visited. The topographic configuration seemingly favors glaciation in this area. The Collanes sector of El Altar was also studied in some detail during the May 1978 field trip. Below the rock precipice of the caldera exit, a flat, swampy basin extends westward for some 3 km, where the valley becomes V-shaped and narrow. On the Southern side of the valley, four moraine terraces (Photo 22) can be traced from below the caldera threshold to the exit of the swampy basin; the higher terraces reaching even beyond this point into the narrow part of the Collanes Valley. Corollaries of these lateral moraines are apparent on the Northern side of the valley at corresponding elevations, although mostly in weaker development. Elevations were measured in the field by aneroid altimeter. The uppermost, largest moraine terrace on the valley slopes can be followed from about 4100 m below the exit of the caldera to little over 3700 m in the narrow part of the Collanes Valley. What appear as terraces when viewed from the valley basin, are upon in situ inspection on the valley slopes found to be distinct ridges. All of these moraines aligned along the valley slopes and in the down-valley part of the swampy basin are covered with a thick layer of volcanic ash. They are strewn with large boulders and carry abundant grass vegetation (Photo 23). These moraines are regarded as belonging to complex III. No volcanic ash was found in the swampy basin, except for the portion down valley from the innermost closed end moraine arc. Evidently the ash was deposited after the moraines had formed, but it should remain open whether the ice still occupied the basin at that time.

Of particular interest is the moraine pair descending from some 4100 m to less than 3900 m immediately below the rock precipice at the caldera exit (Photos 24-28). These moraines are considered to belong to complex II. Most unusual for this elevation range in the Ecuadorian Andes, large polylepis trees with diameters of up to some 50 cm stock on these moraines. The accounts of Meyer (1907) for 1903, of Whymper (1892) for 1880, and of Reiss (Dietzel 1921) for 1870-74 already call attention to the lush tree vegetation. The botanical and meso-climatic implications of this phenomenon remain to be explored. However, a circumstance of key importance for the moraine chronology could be established safely in the field: this moraine pair lacks a volcanic ash cover. It is noted that the deposits of volcanic ash generally appear to favor the vegetation in the Ecuadorian Andes. This moraine pair then is younger than the ash fall covering the lower-reaching moraines of complex III. A further bracketing for the age of this moraine pair is possible from firm historical documentation.

Glacier (2) is referred to as Caldera Glacier by Meyer (1907). Photos

24-28 compare the ice conditions of the Caldera Glacier (2) in 1872, 1902, 1903 and 1978. Photo 24 is a reproduction of Rafael Troya's painting; Photo 25 was taken by Paul Grosser in 1902; and Photo 26 represents a painting by R. Reschreiter during Hans Meyer's expedition in 1903. These Photos are reproduced from Meyer (1907: Vol. 1, Fig. 38; Vol. 2, Plate 18A, B). Photos 27-28 were shot in 1978 from a similar vantage point. Consistent with the text descriptions by Reiss & Stübel (Stübel 1897; Reiss & Stübel 1892-98; Dietzel 1921), Whymper (1892) and Meyer (1907), the photos illustrate the drastic ice recession in the course of the past more than one hundred years. Photo 24 for 1872 shows the ice falling off and accumulating below the large rock precipice. Photos 25 and 26 for 1902-03 depict the ice extending only to the rock threshold. And in 1978 Photos 27-28 indicate that the glacier does not any more reach the exit of the caldera.

Reiss & Stübel (Stübel 1897; Reiss & Stübel 1892-98; Dietzel 1921) studied the area in the course of their 1870-74 expedition (Appendix II: El Altar 1, 2, 3, 4). They describe avalanches crashing down from all sides to the caldera bottom, thus feeding a large glacier; this in turn descending as a partly continuous ice body over a high and steep rock cliff into the swampy Collanes Valley (Photo 24). The ice rim is described as widely separated from a set of large moraines; and the elevation of the glacier snout is set at around 4000 m. The ice thickness at the caldera threshold is estimated at 60-100 m. Whymper (1892) gives a similar description from his visit in 1880, less than a decade later (Appendix II: El Altar 5, 6). He estimated the ice thickness in the caldera at 'a few hundred feet'.

Meyer (1907) encountered a drastically different situation in 1903 (Appendix II: El Altar 7, 8): the regenerated glacier at the bottom of the ice cliff had disappeared altogether, with the Caldera Glacier ending much higher, at the upper edge of the rock threshold (Photos 25, 26). Comparison of Meyer's description and the historical pictures (Photos 24-28) with observations and altimeter readings in the field during May 1978 indicate that Meyer's height figures are positively in error. Concerning the ice thickness, Meyer suggests about 20 m near the caldera threshold and some 50 m further up glacier in the caldera.

The glacier tongue observed by Reiss & Stübel in 1872 as descending the steep rock face below the caldera threshold then extended to near the aforementioned moraine pair of complex II (Photo 24). The moraines below the small glacier tongue (1) in the Northwest part of the caldera around 4400-4300 m, which also lack an ash mantle, are likewise ascribed to complex II (Photo 29). A painting by R. Reschreiter reproduced as Figure 41 in Meyer (1907) indicates that the ice did not extend beyond these moraine ridges in 1903.

Field observations during May 1978 and air photographs indicate a further drastic ice retreat since the turn of the century. Some distinct moraine ridges are presently found inward from the caldera threshold around 4150 m (Photo 30). These do not have any ash mantle and they are vegetated. Since the ice extended beyond this area in 1903, these moraines (I) must have formed sometime during the 20th century. Enquiries from local residents indicate that three small ponds started to form in the early 1930's above the caldera threshold. At present a large lake, about 400 m wide and 700 m long, extends from the caldera threshold eastward, at a level of somewhat above 4100 m. The ice is still in contact with the lake, but over a distance of some 800 m up valley the glacier is covered by enormous masses of debris. Some rudimentary vegetation has started on top of this material. Only further up the caldera at elevations around 4150 m is the ice visible at the surface. The frequent avalanches are still a most impressive spectacle in the caldera.

Based on the above historical documentation, and altimeter readings and observations in the field, an attempt was made to estimate the ice volume loss of the Caldera Glacier (2) in the course of the past 100 years. The volume loss from around 1870 to around 1900, and from around 1900 to present, is calculated to be each of the order of $5 \times 10^7 m^3$. Consider conditions for the average of the 100 year period from the 1870's to the 1970's, a glacier area of the order of $2 \times 10^6 m^2$, and the bulk of the ice loss to be effected through melting with a latent heat of melting of the order of $25 \times 10^4 J kg^{-1}$. The imbalance of energy input corresponding to this rate of ice shrinkage is found to be of the order of 30 W m^{-2} or 60 cal cm^{-2} day^{-1}. Order of magnitude calculations indicate that a change in surface net radiation of this order could be brought about by changes of around 15 % in surface albedo, or of around one to two tenth in cloudiness. By comparison, the change in the liquid water equivalent of solid precipitation would have to be of the order of 500 mm per year. This is consistent with the conclusions arrived at earlier for the recession of glaciers in East Africa (Hastenrath 1975), to the effect that tropical mountain glaciers are less sensitive to changes in solid precipitation than to those in albedo, cloudiness and temperature. A more refined sensitivity analysis is conceivable but does not seem appropriate here. It should be realized that only very crude estimates were used for the parameters entering into this model calculation, i.e. the 100 year ice volume loss, the glacier area, and the relative contributions to ablation of melting vs. evaporation. Although the long-term variations of the Caldera Glacier and of other Ecuadorian glaciers are indeed large, it is found with the aforementioned reservations that they nonetheless seem to stay within energetically reasonable limits.

Historical documentation is lacking for the outer side of the caldera arc. Four separate ice streams (4-7) can be distinguished from air photographs in the North and East sectors. A particularly large glacier (8) is located on the Southeast, and some smaller tongues (9) descend towards lakes on the South side of the mountain. Glacier names ascertained from local mountaineers are as follows: (4) Canónigo or Cerros Negros; (5) Fraile Grande or Naranjal Chico; (6) Los Frailes or Naranjal Grande; (7) Tabernáculo; (9) Obispo.

The study of El Altar thus indicates that bare moraines of complex II in the Collanes sector were not far from the ice rim around 1870 and that moraines of complex I formed during the 20th century. This is broadly consistent with the conclusion reached for the Ilinizas from the joint evaluation of field observation, recent air photography, and historic accounts.

5.6 TUNGURAHUA

This steep ice-capped volcanic cone (Figure 1 and Map 9) in the moist Eastern Cordillera has attracted numerous mountaineers in recent decades. Only partial air photo coverage, a planimetric sheet, and only partially contoured topographic maps are available. The mountain was not visited, and inference relies on air photographs and consultation with local mountaineers.

A complex (III) of seemingly ash-covered and vegetated moraines is found at some low elevations. This may be the corollary of the lowest moraine complex (III) on other mountains.

Apparently bare moraines of another complex (II) can be distinguished at somewhat higher elevations, this again in analogy to other peaks.

Smaller and obviously bare arcs of moraines (I) can be made out in immediate vicinity of the present ice rim, again a feature paralleled on other ice-capped peaks.

Reiss (Dietzel 1921) climbed the mountain in 1873 (Appendix II: Tungurahua 3), and Dressel (1841-44), Whymper (1802), and Meyer (1907) saw it from afar (Appendix II: General 13). Velasco (1841-44) described Tungurahua as snow-covered around the middle of the 18th century (Appendix II: General 6; Tungurahua 2). However, no inference on recent glacier variations was found possible from expedition reports and historical photographs. The present ice extent as derived from aereal photography is asymmetric, the ice limit being lowest to the Southeast and highest in the Northwest, consistent with Stübel's (1897) account (Appendix II: Tungurahua 4) and in similarity to other glaciated peaks especially in the moist Eastern Cordillera.

5.7 CERRO HERMOSO

The Llanganates mountains in the Eastern Cordillera are, exceptionally, not of volcanic origin, but are made up of palaeozoic metamorphic rocks. The highest peak, Cerro Hermoso (Figure 1) carries perennial snow. For the

38

general geology of the region reference can be made to Kennerley (1971) and Kennerley & Bromley (1971). Sauer (1971) shows a photograph of the peak region. I did not visit the area. Air photographs of both 1956 and 1977 and a topographic map are available (Appendix I). No map is presented here because the air photographs do not allow satisfactory identification of moraine morphology and recent ice extent without field visit. In analogy to Soroche & Colay (section 5.2) and Cubillín (section 5.4), white surfaces are smaller on the 1977 than on the 1956 air photographs. Again, secular ice recession and short-term variations in snow cover cannot be safely separated.

Spruce (1861) in 1857 reports only one mountain in the Llanganates as carrying perennial snow (Appendix II: Cerro Hermoso 1). Reiss (Dietzel 1921), Reiss & Stübel (1892-98), and Stübel (1897) in 1870-74 observed a large firn reservoir and glaciers (Appendix II: Cerro Hermoso 2, 3, 4). Wolf (1892) mentions Cerro Hermoso among the peaks with perennial snow (Appendix II: General 21, 22). Andrade Marín (1936) describes the mountain as lacking perennial snow in 1934 (Appendix II: Cerro Hermoso 7). Most recently, Kennerley & Bromley (1971) set the permanent snow line on Cerro Hermoso at about 4350 m (Appendix II: Cerro Hermoso 8). The apparent inconsistency of Andrade Marín's account with earlier and later reports may be related to the particular orographic conditions illustrated in Sauer's (1971) photograph.

5.8 QUILINDAÑA

Concomitant upon its location in the moist Eastern Cordillera, the pyramid of Quilindaña (Figure 1 and Map 10) still carries small glaciers, despite its moderate elevation. The Northern slope of the mountain was visited in December 1974. Persistent clouds associated with severe weather did not warrant continuation of ascent to the peak region.

Moraines (III) of the order of tens of metres high and covered by a several metre thick mantle of volcanic ash and abundant vegetation extend to below 3800 m on the North, and rather lower on the South side of Quilindaña. Distinctly different moraine phases (III) are again apparent. Photo 31 shows a system of moraines (III) enclosing swampy terrain, on the North side of the mountain at 4000 m.

Bare moraine ridges (II and/or I), without ash and vegetation cover, can be identified from the air photographs down to below 4400 m in the West, to around 4200 m in the Northeast, and even somewhat lower to the Southeast of the pyramid. From appearance and elevation these seem to correspond to moraine complex II at the Ilinizas, Corazón and Pichincha. Possible corollaries to complex I could be made out at 4600-4400 m in the Northeast sector of Quilindaña.

Presently, small glaciers exist to the Southeast (1) and the Northeast (2). Snow allegedly persists throughout the year in a depression below steep walls to the Northeast of the peak.

For the middle of the 18th century, Velasco (1841-44) did not seem to consider Quilindaña among the mountains with perennial snow (Appendix II: General 6). Humboldt (1810) states for 1802 that Quilindaña reaches into the region of perennial snow (Appendix II: General 7). Reiss & Stübel (Stübel 1897, Reiss & Stübel 1892-98) studied the mountain extensively in 1870-74 (Appendix II: Quilindaña 3, 4). They mention a hanging glacier above the Toruno-huaico, a cirque to the Northeast of the peak, with falling ice masses accumulating at the valley bottom — a description not unlike Meyer's (1907) account for 1903 (Appendix II: Quilindaña 6, 7, 8), and the present appearance. Dressel (1877) lists Quilindaña among the mountains with eternal snow (Appendix II: General 17). Meyer also mentions small glaciers on the North-northwest and Northwest sides of Quilindaña. The overall impression is one of small hanging glaciers with falling ice accumulating in lower-lying depressions, as presently. Historical accounts and photographs did not allow greater detail concerning secular variations.

5.9 COTOPAXI

The nearly perfectly conical ice-clad mountain (Figure 1, Map 11 and Photo 32) is renowned as the World's highest active volcano. Accounts of disastrous eruptions in historical times have been given by Whymper (1892), Meyer (1907) and others. The reconstruction of climatic history is hampered by the modification or complete obliteration of fossil glacial morphology. In fact, the potential of geomorphic inference at Cotopaxi appears limited. Even the summit elevation of Cotopaxi has allegedly varied in recent centuries (Appendix II: Cotopaxi 1, 2). The usefulness of tephro-stratigraphy remains to be seen. All sectors of the mountain were observed from a distance during the December 1974-January 1975 and May-June 1975 field seasons. The North slope was visited up to around 5400 m.

Possible moraine ridges are apparent on air photographs and in the terrain down to around 4000 m, in the South and Southeast sectors. These ridges are made up of volcanic ash and are essentially without vegetation. These ridges may largely correspond to the lower moraine complex (III) on other mountains, but their nature is dubious in view of the intense volcanic activity.

Moraine-like ridges extending to below 4400 m on the North and South flanks may correspond to the complex of bare moraines (II) on other mountains. The ridges on the North slope in particular are upon inspection in the terrain considered to be of glacial origin. Moraines (I) around 4500 m, little below the present ice rim, are apparent on air photographs on the East and Southeast flanks of the cone.

While the ice covers the volcanic cone as a rather uniform cap, various glacier entities can be distinguished. Thus, several well developed ice lobes are apparent in various sectors of the mountain, partly in close topographic consistency with neighboring moraine-like ash ridges further down valley. The ice cone is distinctly asymmetric, in that the ice reaches farthest down in the Southeast, and has its highest limit in the Northwest, a pattern already noted by Meyer (1907) in 1903 (Appendix II: Cotopaxi 4). Some historic photography is available from Meyer (1907), including the reproduction of a painting from the Reiss & Stübel expedition in 1870-74. Reiss (Dietzel 1921) also gives a figure for the snowline in 1874 (Appendix II: Cotopaxi 3). However, the continued volcanic activity makes attempts at a climatic interpretation ambiguous.

5.10 RUMIÑAHUI

The mountain is here discussed in the context of the Eastern Cordillera, although it more properly lies in the Eastern part of the Interandean Depression. Adjacent to the West but more than 200 m lower than Sincholagua, Rumiñahui (Figure 1 and Map 12) does not take part in the present glaciation. The remnant of the caldera opens towards the Northwest. The mountain was observed from afar in all sectors except the Northeast, but not climbed. Interpretation of glacial morphology relies on air photographs.

Seemingly vegetated moraines (III) extend down to around 3600 m, or near the valley bottom, on the South side, and apparently to much lower elevations in the Northwest. Nesting of moraines is again apparent.

Apparently bare moraine-like features can be identified on air photographs at elevation around 4400-4300 m. It is surmised that these may correspond to moraines of complex II on other mountains. Of the presently unglaciated mountains, corollaries could be detected on Corazón and Pichincha, but not on Atacazo and Pasochoa, which are distinctly lower than Rumiñahui.

For the middle of the 18th century, Velasco (1841-44) lists Rumiñahui among the mountains with perennial snow (Appendix II: General 6). Humboldt's (1810) reference to the mountain suggests that he may have regarded it as not reaching the region of perennial snow (Appendix II: General 7). Definitive statements to this effect are contained in the accounts of Villavicencio (1858) in 1858, Reiss (Dietzel 1921), Reiss & Stübel (1892-98) in 1870-74, Dressel (1877) in 1877, Karsten (1886) in the 1880's, Wolf (1892) around 1890, and Meyer (1907) in 1903 (Appendix II: General 8, 17, 20, 23; Rumiñahui 4, 5, 8).

5.11 SINCHOLAGUA

This mountain (Figure 1 and Map 13) is some 100 m higher than Quilindaña in the same Eastern Cordillera further to the South, and it also participates in the present glaciation. The Northeast and South sides of the mountain were observed from afar, and the Western sector was visited in June 1975 up to around 4200 m. While visibility was fair, the onset of strong Easterly winds at this time of year made a continuation of the trip to the very peak region unsafe.

Large moraine ridges (III), again with a height of tens of metres, and thick ash mantle and abundant vegetation, extend far down the mountain slope to near the valley bottom. Their lowest limit is around 4000 m to the South, below 3800 m to the West, and near 3600 m or lower in the Northern sector. A nesting of different moraine phases (III) is also apparent here.

Bare moraine ridges (II) apparently without ash and vegetation cover can be inferred from the air photographs on the North and South side of the highest crest, at elevations around 4100-4400 m. In appearance and elevation these may correspond to the moraine complex (II) on other mountains. In the Southwest sector of the peak region, which was well visible during the June 1975 field trip, no moraines could be detected in this elevation range. This is consistent with air photography.

Distinct moraine (I) arcs were observed, however, in this sector at an estimated elevation of 4800-4700 m, somewhat below an ice lobe. Terrain observations were verified on air photographs. These likewise show small moraines somewhat below the present ice rim to the North of the highest crest. In context, these moraines seem to correspond to the highest complex (I) on other mountains.

The 'sierra nevada de pinta' mentioned in the municipal records of Quito (González Rumazo 1934a, b) in 1540 and 1550 (Appendix II: General 2, 4) may include Sincholagua and the mountains to the East of Pintag. Contemporary accounts (González Suarez 1890-1903; Bonifaz 1971) attest to abundant snowfall and cold weather in the Eastern Cordillera (Appendix II: General 3). Thus, rather severe ice and snow conditions are suggested for the Pintag region, although a detailed reconstruction is precluded by the scarcity of place names.

For the middle of the 18th century, Velasco (1841-44) lists Sincholagua among the mountains with perennial snow (Appendix II: General 6). Stübel (1897) gives a figure for the snowline elevation on the North side of the mountain in 1870-74 (Appendix II: Sincholagua 5, 6). Dressel (1877) lists the mountain has having perennial snow in the 1870's (Appendix II: General 17).

Whymper (1892) offers a vivid description of snow and ice conditions on the Southeast side of Sincholagua in 1880 (Appendix II: Sincholagua 8). The

Figure 17. View of the Southwest side of Sincholagua in 1880 (from Whymper, 1892).

accompanying etching is reproduced as Figure 17. The peak was viewed from a similar direction during the June 1975 visit. Whymper's picture suggests a vastly larger ice extent in this sector, with a glacier tongue presumably reaching close to the innermost moraine arc apparent in the field and on air photographs. While limitations of drawings as opposed to photographs are recognized, it is considered from this source that moraines of complex (II) were near the ice rim in the latter part of the past century. Aside from the ice lobe (3) in the Southwest sector, one small glacier or perennial ice field (1) is indicated to the North of the crest and one (2) in the South sector.

5.12 PASOCHOA

This mountain is discussed in conjunction with the Eastern Cordillera, although — like Rumiñahui — it more properly lies in the Eastern part of the Interandean Depression. Pasochoa (Figure 1 and Map 14) is also of volcanic origin, and similar to Corazón, Atacazo, and Rumiñahui, the remnant of the caldera opens towards the Northwest. However, of consequence to the glacial relief, Pasochoa is appreciably lower. The West side of the mountain was observed from afar, and geomorphic interpretation relies on air photography.

Seemingly vegetated moraines (III) extend to around 3300 m on the South side, but appreciably lower to the Northwest, similar to Corazón, Atacazo and Rumiñahui, volcanic peaks with a similar crest configuration. Some nesting of moraines (III) is also apparent.

Bare moraines — that is without ash and vegetation cover — which could be regarded as corollaries of moraine complex II on other mountains, could not be detected. This is consistent with the low elevation of Pasochoa as compared to neighboring Atacazo, where evidence for moraines of complex II are likewise lacking. Humboldt (1874a) for 1802, Villavicencio (1858) for 1858, Dressel (1877) for 1870's, Karsten (1886) for the 1880's, and Meyer (1907) for 1903, consistently describe Pasochoa as not reaching the region of perennial snow (Appendix II: General 10, 11, 17, 20, 23).

5.13 ANTISANA

Lying as it does in the moist Eastern Cordillera and looming to more than 5700 m, the Antisana (Figure 1 and Map 15) is at present strongly glaciated, and has a correspondingly abundant fossil moraine morphology. The mountain was visited in June 1974, with field work concentrating on the more easily accessible West sector. Within logistic limitations, an excursion was undertaken along the Southern side of the mountain.

Moraines (III) of a height of tens of metres, and with ash mantle and vegetation cover, extend down to around 4400 m on the West side, somewhat lower to the South, and even to 3400-3200 m in the Eastern sector. A nesting of moraines (III) is pronounced. Major moraine ridges are arranged in spatial consistency with the present glaciers.

Large bare moraines (II) — without ash and vegetation cover — are found nested into the aforementioned complex (III) of ash-covered and vegetated moraines. The striking contrast in the appearance of moraines has been noted by Meyer (1907). The bare moraines of this complex (II) are found at elevations of 4700-4600 m on the West side, but down to near 4000 m in the South and East sectors.

Two particularly large sets of nested moraines are found below a large glacier in the Southeastern sector of Antisana. The outer, lower-reaching set is apparently vegetated and would correspond to moraine complex III in other parts of the mountain. The inner, less extended complex appears to be bare on air photographs and may belong to the aforementioned complex II of bare moraines. Verification in the terrain did not materialize due to logistic difficulties.

A further complex (I) of moraines of fresh appearance — without ash and vegetation cover — is found only tens of metres below the present ice rim, at elevations around 4800 m on the West side, and appreciably lower in the Southeast and East sectors of the mountain. In appearance and altitudinal arrangement this seems to correspond to moraines of complex (I) on other presently glaciated mountains.

Photos 33-41 illustrate the appearance of moraines of complex III, II and I in different portions of the mountain. Photos 33-35, in particular, show the spatial arrangement of the three moraine systems in relation to the present ice rim, for glacier 12 (Great West Glacier; Los Crespos). Photos 36-38 show a closeup of the highest fresh moraines immediately below the present ice rim, in the same area. Comparison of Photos 37 and 38, in particular, illustrates the formation of a moraine ridge since the turn of the century. The corresponding moraine of glacier 14 (Guagraialina) appears on Photo 39. A vegetated moraine of complex III below glacier 10 is seen on Photo 40. The closeup Photo 41 of glacier 10 shows a large bare moraine arc immediately below the snout.

Numerous large glaciers descend from Antisana in all sectors of the mountain. Of these, two were baptized by Meyer (1907) in 1903: Great West (12) and Guagraialina (14) Glaciers. A current name 'Los Crespos' seems to refer to glacier 12. The name 'Glaciar Occidental' seems to encompass glaciers (2), (3) and possibly (4). The term 'Azufral' may refer to glaciers (6) and (7). The currently used name 'Los Cimarrones' presumably refers to glacier (8). Glacier snouts extend to below 4400 m in the Southeastern and Eastern sectors of the mountain, but only to around 4800 m in the West, in similarity to other presently glaciated mountains.

An impression of the present ice conditions is also obtained from Photos 33-41. Photos 33-38 show various aspects of the snout region of glacier 12 (Great West Glacier; Los Crespos). All Photos except 34 and 38 are from June 1975. Meyer's pictures, Photos 34 and 38, taken at the same site in 1903 allow a comparison with conditions around the turn of the century. Comparison of Photos 33 and 34 illustrates the drastic recession of the glacier snout; and Photos 37 and 38 indicate a thinning of the lower glacier since the turn of the century. Photo 39 shows the ice towers forming the end of glacier 14 (Guagraialina); and Photos 40-41 illustrate the area of glacier 10 in the Southwest sector.

Reiss & Stübel (Stübel 1897; Reiss & Stübel 1892-98) studied the mountain in 1870-74, but their references did not contribute to climatic reconstruction. Whymper's (1892) narrative is likewise not substantial on ice rim conditions. Meyer (1907) in 1903 approached the mountain from the West, where he baptised two glaciers (Appendix II: Antisana 1). Some inference on glacier behavior since the turn of the century is possible with reference to his photographic documentation. For glacier 12 (Great West Glacier; Los Crespos), it was possible to identify in the terrain the location from which Meyer's photograph was taken at the beginning of the century. Photography was repeated at the same location in June 1975 (in particular Photo 37). In conjunction with Meyer's picture (Photo 38), a drastic ice recession is borne out since the beginning of the century, although any intermediate variations or stages cannot be reconstructed. From a comparison of photographs it can furthermore be concluded that the ice rim must have been close to the afore-mentioned uppermost bare moraines (I), at the time of Meyer's visit. Comparison of photographs also suggests a thinning of glacier 12 (Great West Glacier; Los Crespos) by tens of metres since the beginning of the century.

5.14 SARA URCU

This mountain (Figure 1) is of rather modest elevation as Ecuadorian summits go, but lying on the wet Amazonian side of the Andes it is strongly glaciated. Visits to this remote region are rare. Air photographs are available of 1966 and 1978 (Appendix I). In view of its far eastward location, exploration of fossil and recent glaciation of this mountain would have particular merit. References to the snow and ice conditions on Sara Urcu (Appendix II: General 6, 19, 22; Sara Urcu 2-5) are available from the middle of the 18th century (Velasco 1841-44), from 1870-74 (Reiss & Stübel 1970-74; Stübel 1897; Dietzel 1921), 1880 (Whymper 1892) and 1892 (Wolf 1892) – with Whymper's book even containing an etching of the peak region – but these do not allow a reconstruction of secular variations.

5.15 CAYAMBE

Rising to more than 5700 m in the moist Eastern Cordillera, Cayambe (Figure 1 and Map 16) is among the most strongly glaciated mountains in the Ecuadorian Andes. Travel reports from earlier centuries are limited. Reiss and Stübel visited Cayambe in 1870-74, Whymper climbed it in 1880, and Meyer studied the mountain in 1903. Some photographic documentation is available from subsequent decades. Partial air photo coverage and planimetric maps exist for the area, but no contoured topographic sheets. The South side of the mountain was visited during June 1975. The limited field observations are evaluated in conjunction with the air photographs.

Large moraines (III) with a height of the order of tens of metres extend down to elevations around 3700 m in the sector visited. Photo 42 shows a moraine of this complex on the South side of the mountain, with the ice rim in the more distant background. The moraine ridges carry a thick mantle of volcanic ashes and are vegetated. The air photographs show corresponding moraine features in other sectors of the mountain.

Moraines (III) apparently without ash cover and vegetation are found at somewhat higher elevations, in similarity to other mountains. Another complex of seemingly fresh moraine (I) arcs is found in the vicinity of the present ice rim, again in analogy to other glaciated mountains.

Air photographs suggest some asymmetry in the present ice distribution also for Cayambe with lowest snouts in the Eastern sector and higher elevations of the ice limit to the West. This appears broadly consistent with observations by Reiss & Stübel (1892-98) and Stübel (1897) in 1870-74 (Appendix II: Cayambe 1, 2, 3). Whymper's (1892) narrative for 1880 (Appendix II: Cayambe 4) is less informative. Names ascertained from local mountaineers are as follows: (17) Espinoza; (20) Chuquira Cucho. A quantitative comparison of the accounts of Reiss and Stübel with present conditions may have promise. A secular recession is suggested by the fresh moraine arcs in the immediate vicinity of the ice rim.

6 OLDER GLACIATIONS

Soweit heute unsere Erfahrungen reichen, scheinen
mir keine zwingenden Gründe für die Annahme einer,
auch die Cordilleren Ecuadors umfassenden, allge-
meinen Eiszeit vorzuliegen.

Wilhelm Reiss, *Das Hochgebirge der Republik Ecuador,*
1892-98.

At least three distinct glacial events can be inferred from moraine complexes
arranged in spatial consistency with the present glaciers. The oldest and most
extensive of these glaciations reached down to less than 3500 m. Meyer
(1907) and Sauer (1965, 1971) postulated earlier and much lower reaching
glaciations largely from evidence other than moraine morphology. Epirogenic
uplift of the Andes is regarded as small during the recent geological past
(Petersen 1958), and may thus not exclude the possibility of earlier glacial
events. In fact, Herd & Naeser (1974) presented evidence for a pre-Wisconsin
glaciation of the Northern Andes.

Some of the type sites described by Sauer were visited in May-June 1975.
The Rio Chiche profile along the Quito-Pifo road at around 3000 m (Figure
1) was located in all likelihood. However, Sauer's interpretation of strata as
manifesting distinct glacial events could not be shared as unambiguous. As
further evidence Sauer describes a profile near the electric plant of Guangopolo at 2400 m (Figure 1), containing presumably glacially transported
debris material with large rounded rocks. The profile was located beyond
doubt. Interpretation in terms of glacial origin is considered plausible. This
would imply glacial transport to elevations well below the ubiquitous ash-
covered moraine ridges.

Large, nearly horizontal terraces at two to three distinctly different levels
were observed in valleys on the West side of Iliniza (Photo 43) and the North
side of Atacazo. Cuts show deep beds of coarse gravel. These terraces are
reminiscent of the ones described by Meyer (1907) and Sauer (1965, 1971)

48

as evidence of earlier glaciations. A relation to glacial events in the Western Cordillera seems plausible, but a conclusive spatial and chronological correlation is lacking.

More definite indication of possible earlier glaciation is provided by roches moutonnées and glacial striations described by Sauer (1971) for comparatively low locations. Photo 44 shows a roche moutonnée with striations of seemingly glacial origin in the village of Papallacta (Figure 1) on the Eastern side of the Eastern Cordillera at 3200 m. Similar striations were found at other locations in the Papallacta valley down to around 3000 m, but an associated moraine morphology could not be identified from field observation nor air photographs. Striations are oriented approximately NNE-SSW, broadly following the large-scale topography. The plausible catchment area for the corresponding ice stream is from the regional topography found to be at elevations of only 4000-4500 m. A travel to the lowlands of the Rio Napo in the Oriente during June 1979 offered the opportunity to reconnoitre the entire Papallacta valley to well below 2000 m. No moraine morphology could be detected.

Sauer mentioned glacial morphology at very low elevations in the Cuenca region, but these type sites were not visited. He reported a terminal moraine at 1750 m and explained glacial traces as low as 800 m by subsequent tectonic activity.

In the mountains to the West of Cuenca deposits without distinct surface morphology have been mapped as moraines at elevations as low as 1800 m (Ing. Jorge Guzmán, Quito, personal communication 1978). It remains open whether these features correspond to the moraines of complex III in other parts of the Ecuadorian Andes or to an older glacial event. By way of spatial comparison, the terrain should certainly be considered as a factor. Elevation in this area ranges, over a short distance, from more than 4000 m to well under 2000 m. Surfaces of such a low elevation are not commonly encountered in the vicinity of the glaciated peaks in the two cordilleras.

In context the field evidence from the Papallacta and Guangopolo sites and the Cuenca region points towards the possibility of at least one glaciation that was older and lower-reaching than the glacial events manifested by the three moraine complexes (III, II, I) in the High Andes.

7 SUBNIVAL SOIL FORMS

... desde donde ya empieza à mantenerse la Nieve
algun tiempo, sin derretirse, no crece ninguna de las
Plantas, que son regulares en los Climas habitables; pero
sì otras en su lugar, aunque raras, hasta una cierta altura;
desde la qual en adelante no se encuentra mas, que
Arena, y Piedras por larga distancia, hasta llegar al
principio de la congelacion.

Jorge Juan and Antonio de Ulloa, *Relacion historica
del viage a la America Meridional,* 1748.

The altitudinal zonation of periglacial phenomena is related to the temperature characteristics described in Chapter 3. Consider a mean annual elevation
of the $0°C$ isothermal surface somewhat below 4900 m, annual and daily
temperature ranges of about 2 and $4°C$, respectively, and a climatic mean
lapse rate of the order of $0.65°C/100$ m. Then nighttime frost and daytime
thawing can be expected with some regularity upward of about 4300 m. A
variety of subnival soil features is in turn related to the change of frost.

Needle ice is found commonly above about 4300 m, in agreement with
the aforementioned temperature conditions. The fine earth remnants of
needle ice tend to be arranged in approximately East-West oriented stripes,
following the local sunrise, a phenomenon known from other mountain
regions of the tropics. The destructive effect of needle ice on the vegetation
cover is obvious.

A variety of other soil frost phenomena occur frequently in the altitude
range of 4500-5000 m. These include fine earth polygons, stripes in fine
material (Photo 45), stone polygon (Photo 46) and stripe patterns (Photo
47).

All of these features have been described from other high mountain
regions of the tropics. However, remarkable in the Ecuadorian Andes is the
comparatively rare occurrence of all subnival soil forms. This appears broadly

Photo 39. Ice towers forming end of glacier 14 (Guagraialina) at 4800 m, with fresh moraine in foreground (June 1975).

Photo 40. Antisana from the Southwest at 4200 m. To left vegetated moraine of complex III; in center glacier 10.

Photo 41. Antisana from the Southwest at 4200 m. Same position as Photo 35, but telephoto of glacier 10.

Photo 42. South side of Cayambe at 4000 m. Vegetated moraine (III) in foreground; ice rim in background (June 1975).

Photo 43. Terraces on West side of Iliniza at 3200 m (Dec. 1974).

Photo 44. Roche moutonnée with glacier striations; Papallacta, 3200 m (June 1975).

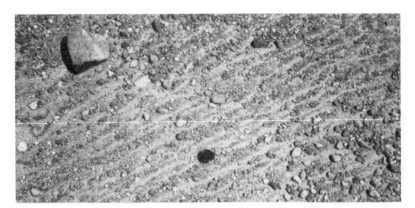

Photo 45. Stripe pattern in fine material. South side of Carihuairazo at 4300 m. Disc as scale has diameter of 7 cm (Dec. 1974).

Photo 46. Stone polygons, South side of Carihuairazo at 4300 m. Box as scale has diameter of about 7 cm (Dec. 1974).

Photo 47. Stone stripe pattern on North side of South Iliniza, at 4700 m. Disc as scale has diameter of 7 cm (Dec. 1974).

compatible with observations of Graf (1976) and Furrer & Graf (1978). It is surmised that the impoverishment of periglacial morphology is related to the ubiquitousness of volcanic ashes, uniformity in grain size distribution not being conducive to pronounced material sorting. A somewhat similar state of affairs has been described by Schunke (1975) for Iceland.

8 CONCLUSIONS

... llega hombre fasta donde puede y no fasta donde
quiere ...

Vasco Nuñez de Balboa, *Letter to the King of Spain,*
20 January 1513 (Archivo de Indias).

Evidence is presented in Chapters 4 to 6 (summary in Tables 2 and 3 and
Figure 18) jointly for the recent and former glaciations, because glacier
systems are regarded as temporally continuous entities. Also, a chronology
cannot be prescribed a priori, but must be deduced through the combination
of field observation, air photo interpretation, and evaluation of historical
sources. The presently glaciated peaks attract attention in the first place,
since the relation to fossil moraine morphology may seem more obvious.
However, several mountains of moderate elevation in the vicinity of Quito
which presently lack perennial snow, have proven especially informative for
glacial reconstruction: these peaks have been repeatedly mentioned by early
travellers, and the presence or absence of snow is definitive information,
whereas early figures on the elevation of glacier snouts are intrinsically
uncertain. An attempt has been made to take stock in the glaciation of the
Ecuadorian Andes as a whole, with the premise that spatial patterns and
prominent glacial-climatic events would emerge more clearly than could be
expected from the more detailed study of a single mountain massif. A trade-
off of this nature was also dictated by limitations in weather conditions,
accessibility, and logistics.

8.1 SPATIAL PATTERN OF GLACIATION AND MORAINE CHRONOLOGY

Inevitably the inventory of present ice extent is incomplete, although an
attempt has been made to identify individual glacier tongues on the various

mountains. Among the prominent features of the large-scale pattern are the lower ice equilibrium limit in the Western as compared to the Eastern Cordillera, and the greater ice extent in the Eastern and in part the Southern sectors as opposed to the Western portion of individual mountains. These features have been recognized by early explorers, notably Reiss and Stübel (Reiss 1897; Reiss & Stübel 1892-98; Dietzel 1921), and Meyer (1907). By contrast, a more intense glaciation in the Western than the Eastern quadrants has been reported by Sievers (1914) for the neighboring Andes of Northern Peru. A similar state of affairs is found in the Northern hemisphere portion of the American Cordilleras, such as on the Mexican volcanoes (Lorenzo 1964; Heine 1975), as well as in other parts of the Tropics (Hastenrath 1974b, 1975, 1977b; Nogami 1976).

A pronounced afternoon maximum in the diurnal march of cloudiness would reduce insolation and favor glaciation to the West. This effect associated with the diurnal cycle may be of subordinate importance in the chronically cloudy environment of the Ecuadorian Andes. The stronger glaciation in the Eastern Cordillera, and in the Eastern sectors of individual mountains may reflect the moisture supply from the Amazon basin. Remarkably, the same East-West asymmetry is also apparent in the Western Cordillera, especially in the glaciation of Chimborazo-Carihuairazo. With regard to moisture sources, it is recalled (Chapter 3) that an altitudinal belt of abundant rainfall extends along the Pacific slope of the Andes, and a shallow clockwise turning cross-equatorial surface flow persists throughout the year over the Eastern Pacific.

It should be noted, however, that moraine-like features possibly belonging to complex III are found at exceptionally low elevations in the Western quadrants of certain mountains, namely the calderas which as a rule are open in that direction, such as Corazón, Atacazo, Pichincha, Rumiñahui and Pasochoa. Whether the aforementioned systematic asymmetry in the rim of calderas is related to the prevailing Easterly winds or glacial erosion in the first place, remains open. In any case, the wide caldera basin would be most favorable for ice accumulation. Formerly low-reaching glaciers on the Western side of presently unglaciated caldera mountains may be related to this factor and not necessarily reflect a change in the spatial pattern of atmospheric circulation and climate (Chapter 3). In fact, the only presently glaciated caldera mountain, El Altar, has on its Western side the apparently lowest-reaching 'live' glacier in the Ecuadorian Andes.

As a further factor for the elevation of glacier snouts, the gross topography should be taken into account: thus ash-covered and vegetated moraines (III) extend to rather low elevations on the North side of Sincholagua and Rumiñahui – whereas the valley to the South has a much higher floor; similarly, the plateau to the West of Chimborazo-Carihuairazo is much higher than the Eastern foot of these mountains, where the lowest-reaching moraines are found; and Iliniza, Corazón, Atacazo and Pichincha, drop off sharply towards the Pacific coast. The gross topography, and the existence of low-lying surfaces near possible centers of former glaciation, along with the abundance of precipitation, may also be factors for the low-reaching moraines reported in the Cuenca and Papallacta regions.

The correlation of moraines on the various mountains must rely on spatial arrangement, altitudinal range, and appearance. Absolute moraine dates are not available, and from the scarceness of vegetation in the high regions it seems unlikely that C14 datable material can be found in situ. A large-scale tephro-stratigraphy may have potential, however, in that spatially continuous ash layers may overlie both the moraines in the mountains and remnants of vegetation in the lower regions of the interior basins and valleys. Organic material at the base of the volcanic Tarqui formation in the Azogues area was dated at 24,900 and 34,300 B.P. The Tarqui formation in turn underlies the complex of lowest-reaching, ash-covered moraines (III) in the Cerro Rasullana region (section 5.1). Inasmuch as appreciable tolerance must be allowed for the age of Tarqui deposits at different locations, the figure of 25,000 B.P. has only limited validity as a maximum age of the complex III moraines. The timing of ash deposition — and various phases are to be distinguished here — would provide a maximum age for the lowest-reaching, ash-covered moraine complex III, and a maximum age for the next higher complex II of moraines without ash mantle and vegetation. In this connection it should be noted that the continuous layer of volcanic ash gradually vanishes towards the South of Ecuador.

Barring absolute dates, the lowest, ash-covered and vegetated moraines at elevations from more than 4600 to less than 3500 m have been regarded as belonging to a synchronous glacial event on all mountains, and labelled complex III. It is recalled, however, that there is appreciable spatial detail to this complex: at least three distinct, parallel moraine ridges can be distinguished in places, with further moraine arcs at somewhat higher elevations, such as on the East side of Chimborazo. The regularity of these features suggests that they may be recessional moraines. On the whole, the lower limit of this moraine complex shows a pattern similar to the present glaciation, in that lowest elevations tend to be reached in the Eastern and Southern sectors of individual mountains, such as on Chimborazo-Carihuairazo.

Moraines of this complex III mark the most extensive and oldest glaciation that can be substantiated in terms of moraine morphology. Evidence has been reviewed in Chapter 6 that may in part pertain to this glacial event, but part of which suggests the possibility of at least one even older and lower-reaching glaciation.

The prospect of estimating the timing of glacial climatic events from archeological evidence is intriguing. Based on age determinations by the hydration rim technique, Bonifaz (1978; personal communication 1978) believes that obsidian artifacts in the 2800-3400 m elevation range in the Cayambe and Ilaló areas appeared essentially after 10,000 B.P. This phenomenon may reflect a climatic event. In view of the considerable tolerance in elevation, however, it seems unwarranted to relate this archaeological evidence to the ice retreat from either the complex III moraines, or alternatively any earlier lower ice maximum.

More palpable is the evidence for glacial events younger than the aforementioned moraines of complex III. In particular, moraines of a distinctly different appearance are found mostly at elevations of 4600-4200 m, on all presently glaciated mountains, but also on some mountains that now stay below the snowline. These moraines are completely bare, that is, they lack ash mantle and vegetation. From appearance, spatial arrangement, and elevation range, corollaries from the various mountains are proposed to belong to the same glacial event, and are referred to as 'complex II'. In some places, such as on the West side of Carihuairazo, the East side of the Ilinizas, and the West side of Antisana, these moraine ridges are located immediately adjacent to and somewhat above those of complex III: at such locations, the contrast between the bare moraines II and the ash-covered and vegetated ridges of complex III is particularly striking. Moraines of complex II can be expected at various presently unglaciated mountains, which I did not visit.

8.2 ICE EQUILIBRIUM LINE IN HISTORICAL TIMES

The schematic diagram Figure 18 constructed from Tables 2 and 3 provides a pictorial synopsis of the historical and fossil geomorphic evidence. There is a systematic difference in snowline elevation between the Western and Eastern Cordilleras although this cannot be substantiated for the earlier epochs. Figure 18 illustrates the gradual upward displacement of the snowline in the Andes near Quito in the course of the past several 100 years. Glaciation during the 16th and the first half of the 18th centuries was considerably more intense than at present. In particular, there are indications that a portion of the Eastern Cordillera to the East of Pintag which rises to little more than 4400 m may have been perennially snow-capped during the 1500's. Moraines of the (lower) complex (II) may have been in contact with the ice around the 1700's or earlier. Moraines of this complex (II) are found on several mountains that still reach the snowline in historic times but not at present. Moraines of the (higher) complex (I) emerge as a feature of the 20th century. In fact, moraines I are found on mountains that are presently still glaciated.

From spatial arrangement and stratigraphy moraines II must belong to a glacier advance younger than the moraines of complex III. Concerning the

Table 2. Glaciation in the Western Cordillera. III, II, and I refer to altitude range of moraine complexes; SL and GL signify estimates of the present ice equilibrium line and glacier limit; and approximate years are given for sources pertaining to secular variation of the recent glaciation, with + and − signs denoting that mountain does or does not carry perennial snow/ice. Asterisks (*) indicate mountains presently glaciated. Elevations in m.

Mountain	Elevation	III	II	I	SL	GL	1500's	1740	1802	1858	1872	1880	1903
Cotacachi*	4939				?	?							
Pichincha	4675	N 4200-3200 S 4300-3400	4600-4500 NW 4400-4200				+	+	+	(±?)	(+?)	(+?)	(?)
Atacazo	4570	E 4200-3800 NW, SE 4100-3300					(+?)		−	(+?)			
Corazón	4788	4000-3200 SE 4200-3800	4600-4400 NW 4300-4000				(+?)	+		+	+	(+?)	(+?)
Iliinizas								SL 4700					
Norte	5126	4300-3400	4700-4200	4800-4750	(>4800?)	around 4800 lowest NW of Iliniza Sur							
Sur*	5248										saddle to W SL 4658 GL 4484		
Carihuairazo*	5020	N 4000-3200 S 4800-3600 W 4200-4000	4400-4200 W 4600-4400	4700-4500	(~4800?)	4700-4500							
Chimborazo*	6310	N 4600-3700 S 4600-3600 W 4500-4200	5000-4400	5000-4600	(>4800?)	5000-4600							

56

Table 3. Glaciation in the Eastern Cordillera. Format and symbols as for Table 2.

Mountain	Elevation	III	II	I	SL	GL	1500's	1740	1802	1858	1872	1880	1892	1903
Cayambe*	5790	S -3700			?	?								
Sara Urcu*	4676				?	?					SL W 4364	+		
Antisana*	5753	W 4600-4400 S 4600-3800 NE 4400-3400	4700-4600 SE 4400-4200	4700-4200		W-SW 4700-4600 SE 4400 NE 4200						+		
S. Pintag Pasochoa	4400+	NW 3600-3000					+						–	
Sincholagua*	4893	4200-3800	4600-4400	SW 4700		<4800					SL N 4577	more than now		
Rumiñahui	4712	4200-3600 NW 3900-3100	–<4400					–		–	–			
Cotopaxi*	5911	4500-3700				SW 4700 SE>4400								
Quilindaña*	4760	4200-3800 SW 4100-3600	4400-4200	N 4600-4400		4600?			+					small gl. NNW, NW
Cerro Hermoso*	4571				4350	?				+				
Tungurahua*	5016				?	?				+	SL 4650 GL W 4000	+	+	
El Altar*	5319				?	>4300				+	GL W 4300	GL W 4300		+ GL W 4300
Sangay*	5230				?	?				(+)				
Rasullana	4475	4200-3600												

57

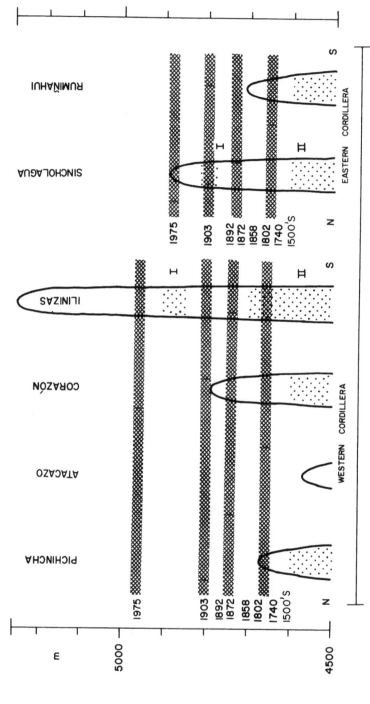

Figure 18. Schematic diagram of snowline variation in the Andes of Quito from the 1500's to present. Cross-hatched bands indicate snowline elevation, and dotting signifies moraine belts (II and I). Data are taken from Table 2.

minimum age of complex II, historical records of the past several centuries are of interest (Tables 2 and 3). For the Ilinizas and El Altar there are eye-witness reports from around 1872 that place the glacier end somewhat above moraine complex II as identifiable from recent field observations and air photographs. Specifically, the glacier snout is given an elevation of 4484 m on Iliniza and of 4000 m at El Altar. Evidence for Sincholagua and El Altar from 1880 is less detailed, but points in the same direction.

These glacier observations for Iliniza, Sincholagua, and El Altar are complemented by reports on the presence or absence of perennial snow on lower mountains along the margins of the Interandean Depression. Thus, Pichincha had a perennial snow/ice cover in the 1500's and maintained it from the first half of the 18th until the early 19th century, after which time the snow cover became marginal; it seems to have disappeared for good by the beginning of the 20th century. By comparison with Pichincha, Corazón must have shared the perennial snow cover in the 1500's. It reportedly remained perennially snow-capped from around 1740 until at least around 1872. Snow and ice conditions became marginal by the beginning of the 20th century. Rumiñahui, Pasochoa and Atacazo, are consistently described as free of snow from 1802 onward.

It is also noted from Tables 2 and 3 that the lower mountains Pasochoa, Rumiñahui and Atacazo lack moraines of complex II. Contrarywise, the somewhat higher mountains Pichincha and Corazón possess moraines of complex II — and they were perennially snow-capped in the 1500's and from around 1740 until different decades of the 19th century. Reports on the snow conditions on these lower mountains along with the observations of glacier snouts on Iliniza, Sincholagua and Altar, indicate that the ice rim was somewhat above the moraines of complex II in the era from around the middle of the 18th to around the middle of the 19th century. Moraines of complex II are thus bracketed as being younger than the ash falls postdating moraines III, but older than the middle of the 19th century.

In context, a maximum age of several 100 years seems plausible for the moraines of complex II. In fact, comparatively severe snow and ice conditions during the 1500's and the 1700's can be inferred from the aforementioned historical accounts in the Ecuadorian Andes. For the region of Ananea in Southeastern Peru there are reports (Pflücker 1905; Sievers 1908; Broggi 1943; Oppenheim & Spann 1946) to the effect that a glacier vacated former mining sites around the turn of the century. The evidence of former mining activity is believed to date back to the time of the Spaniards. The only pertinent field observations appear to be due to Pflücker, however, and comparison of the various articles shows inconsistencies. It seems that these could be resolved only by verification in the terrain.

Certainly as recently as in the second half of the 19th century, the ice extended to elevations well below the altitude range of moraines of the next higher and younger, complex I. These moraines are found only little below the present ice rim. At least in part these may have formed at some time in the course of the 20th century. As corollaries from other tropical regions, small moraine arcs younger than 1920 are found in the Cordillera Blanca of Peru (Kinzl 1949, 1968; Clapperton 1972; Lliboutry et al. 1977), and in East Africa (Hastenrath 1975). Glacier retreat since the turn of the century has been reported for other regions of the South American Andes (Broggi 1943; Oppenheim & Spann 1946; Petersen 1967; Wood 1970; Schubert 1972a).

8.3 VARIATIONS IN ICE-COVERED AREA

Table 4 compares the present ice extent with the approximate ice-covered area corresponding to moraine stages III, II and I. Estimates were derived

Table 4. Approximate ice-covered area (in km^2) corresponding to moraine stages III, II, I, and present (1970's, P). The outermost moraine ridges are considered for III. Parentheses denote very crude estimates.

	III	II	I	P
Western Cordillera				
Cotacachi	37	(2.5)	(1)	0.5
Pichincha	100	1	0	
Atacazo	22	0		
Corazón	51	1.5	0	
Ilinizas	158	61	2	1.5
Carihuairazo	319	15	7	6
Chimborazo		57	37	30
other	(300)	(0)		
Eastern Cordillera				
Cayambe	147	70	(50)	44
Antisana	67	40	30	25
Pasochoa	17	0		
Sincholagua	108	(8.5)	3	(1.5)
Rumiñahui	71	2.5	0	
Cotopaxi	(80)	(50)	(32)	27
Quilindaña	73	(4)	1	(1)
Tungurahua	(60)	(20)	(8)	7
El Altar	178	85	(65)	60
Cerro Rasullana	15	0		
other	(257)	(42)	(19)	(16.5)
sum, Western Cordillera	987	138	47	38
sum, Eastern Cordillera	1063	332	208	182
sum, both cordilleras	2050	460	255	220

from the Maps 1-16 by planimetering of areas delineated by moraines and gross topography. The approximate contribution of other mountain areas was estimated from their elevation and horizontal extent. For the Western Cordillera, the extensive regions above 4000 m in the Southern Provinces of Azuay and El Oro provide a major contribution to the ice extent of stage III. The Eastern Cordillera contains some mountains that are presently glaciated but for which a map could not be constructed for lack of air photographs and topographic sheets. The estimates for mountains in the Eastern Cordillera that are not depicted in Maps 9-16 therefore include contributions for stages III, II, and present.

It is noted from Table 4 that for the Ecuadorian Andes as a whole, the total ice-covered area at the maximum of stage III was about an order of magnitude larger than at present. At stage II the ice was more than double as extended as now, while stage I differed comparatively little from the present conditions.

The breakdown in Table 4 shows a rather different pattern for the two cordilleras. The contrast between the two cordilleras is largest at present, but was much smaller during the earlier stages, especially III. Viewed differently, the decrease in ice extent from stage III to II, I and to present, was most drastic in the Western Cordillera, where environmental conditions seem less favorable for glaciation. By contrast, the Eastern Cordillera, which is throughout characterized by abundant moisture supply from the Amazon basin, and by the more vigorous glaciation, was much less affected by long-term changes.

8.4 SYNOPSIS OF GLACIAL EVENTS IN THE TROPICS

Table 5 offers a synopsis of moraine stages in the high mountains of the tropics; regions being arranged proceeding from North to South, and separately for the Americas, Africa and Australasia. A conclusive spatial correlation is precluded by the scarcity of absolute dates. Arrangement in columns only to suggest plausible corollaries between regions.

In all of the American Tropics, the best established glacial chronology is now available for Mexico (Heine 1975). A plausible correlation offers itself southward into Northern South America, in that the moraine complexes in the mountains of Guatemala and Costa Rica (Hastenrath 1973, 1974a), and the higher moraine complex in the Venezuelan Andes (Schubert 1972a, b, 1974), may correspond to the double complex M III on the Mexican volcanoes (Heine 1975). Some reservations to this proposition are as follows: the upper, small moraines in the Cordillera de Talamanca (Hastenrath 1973) could be younger, and the higher moraine complex in the Venezuelan Andes (Schubert 1972a, b, c) could be older than M III of Mexico. Younger

Table 5. Synopsis of moraine stages in the high mountains of the tropics. H, SL, lat. refer to elevation of highest peak and modern snowline, and latitude, respectively. Ages in parentheses are assumed, not measured.

	H lat.	SL	M V	M IV	M III	M II	M I
a. Americas							
Mexico (Heine 1975)	5670 19 N	(5000)	5100-4300 m 1800+ A.D.	4200-3800 m 2000 B.P. weak	1. 4000-2900 m 9000 B.P. 2. 3500-3000 m 10,000 B.P.	3200-2600 m 12,100 B.P.	2800-2600 m 34,000-32,000 B.P.
Guatemala: Cuchumatanes (Hastenrath 1974a)	3837 15 N	—			3600-3500 m 4 separate ridges		
Costa Rica: Talamanca (Hastenrath 1973)	3819 10 N	—			1. 3500-3380 m up to 3 ridges, small 2. 3350-3300 m 3 separate large ridges		
Venezuela (Schubert 1972a, b, 1974, 1975)	5002 9 N	(4750)		4300-4000 m weak	3700-3000 m >10,000 B.P. or >13,000 B.P.	3000-2600 m	
Colombia Santa Marta (Gansser 1955; Raasveldt 1957)	5775 11 N	(4800)	Bolivariano		Mamancanaca	Aduriameina	
Colombia: Cocuy (Gonzalez et al. 1965)	5490 7 N	4600	4. 4200 m (1700- 1800 A.D.?)	3. 4200 m 2900- 2300 B.P.	2. 4050 m >7500 B.P.	1. 4000 m	
Colombia: Ruiz-Tolima (Herd/Naeser 1974)	5420 5 N	(4600)			3450 m 3 phases <13,760 B.P.		3450 m >100,000 B.P.

62

Region (reference)	Elev. (m) / Lat.						
Ecuador (Hastenrath 1981)	6310 0-5 S	(4900-4300)		I. 4900-4600 m ~1900+ A.D.	II. 4700-4200 m ~1500-1800 A.D.	III. 4500-3400 m 1. upper complex, smaller 2. lower complex, large, up to 3 separate ridges	glacier striations down to <3000 m
Peru: C. Blanca (Clapperton 1972)	6768 8 S	5000		4. 4750-4250 m -1900 A.D.	3. 4650-4200 m (1750-1800 A.D.)	2. 4200-2000 m (6000-4000 B.P.?)	1. 4000-3000 m
(Lliboutry et al. 1977)			Safuna VI 4440-4380 m 1900+ A.D.	Safuna V/ Huaraz m8 4600-4400 m 1,600+ A.D.	Safuna IV-III/ Huaraz m7 4500-4240 m	Safuna II-I/ Huaraz m6-m5 4400-3900 m (7,000 B.P.)	Safuna Huilcapampa/ Huaraz m3-m2 3900-3500 m Safuna Upper Terrace Huaraz m4-m1 3900-3200 m
Peru: Quelccaya/ Vilcanota (Mercer et al. 1975) (Mercer & Palacios 1977)	5645 14 S	5400	5100 m (Q) <270 B.P. 4550 m (V) <630 B.P.	4,500 m(V) >2,830 B.P.	5050 m (Q) <10,900 B.P.	5050 m (Q) >12,240 B.P.	4450 m (V) 28,000-14,000 B.P.
Chile: Elqui Valley (Caviedes & Paskoff 1975)	6252 30 S	5500-5000			La Laguna 3100 m	Tapado 2500 m	
b. Africa Ethiopia: High Semyen (Hastenrath 1974b, 1977c)	4543 13 N	—	—			4000-3750 m	
Ruwenzori (de Heinzelin 1962; Livingstone 1962; Whittow et al 1963; Osmaston 1965; +Hastenrath)	5110 0	(4500)	younger moraines Lac Gris 4400-4600 m 4-6 phases	Lac Noir et Vert/Omuruba-ho 3500-4300 m 2 phases	Kichubu and Bigo 3060 m	Butahu, Lake Mahoma 2300 m deglaciation 14,700 B.P.	older glaciation? <2600 m Ruimi Basin 1890 m Crête Ruamya-Haute Ruanoli Katabarua

Table 5 (continued)

	H, lat.	SL		VI A,B	II-V	I A-D	'older glaciation'
Mt Kenya (Baker 1967; +Hastenrath)	5199, 0	(4800)	younger moraines	≥4600 m, 2 phases (1820–1850 A.D.)	4300–4000 m	'Younger Maxima', 3400 m	'older glaciation'
Kilimanjaro (Downie & Wilkinson 1972; +Hastenrath)	5395	(4800)	younger moraines	'Recent glaciation' 5200–4600 m, 2 phases (1770 A.D. and 2000 B.P.)	'Little glaciation' upper; lower, 4 phases, 4000 m (10,000–8,000 B.P.)	'Fourth (Main) glaciation' 2 phases, 3400 m (30,000–10,000 B.P.)	Third glaciation >100,000 B.P. Second glac. 300,000 B.P. First glaciation 500,000 B.P.
c. Australasia							
Borneo, Kinabalu (Koopmans & Stauffer 1968)	4100, 6 N	–				2800 m	
Irian Jaya: Carstensz (Peterson & Hope 1972; Galloway et al. 1973; Hope et al. 1976)	5029, 4 S	4400		4200 m 3,500 B.P. + three more advances >1,600 B.P.		3500 m <11,300 B.P.	deglaciation 3680 m >14,000 B.P.
Papua New Guinea (Loeffler 1972, 1976, 1979)	4509, 6 S	–				3700 m >10,700 B.P.	deglaciation 3600 m Mt. Wilhelm >12,600 B.P. glaciation 290,000 B.P. glaciation >380,000 B.P.

moraines are lacking from Guatemala into Venezuela, a circumstance probably related to the lower summit elevations. Schubert (1972a, b, 1974) regards the lower moraine complex in the Venezuelan Andes as equivalent to the Mamancanaca stage of the neighboring Sierra Nevada de Santa Marta of Colombia (Gansser 1955; Raasveldt 1957; Bartels 1970). Raasveldt's (1957) description in turn would put the Bolivariano moraines of the Sierra Nevada de Santa Marta into the era of the Mexican MV moraines (Heine 1975).

Correlation of moraine stages from Northern South America southward appears more difficult. Schubert (1972a, b, 1974) envisages an equivalence of Mamancanaca with Clapperton's (1972) moraines of stage 1 in the Cordillera Blanca of Peru, but this shall be left open. Similarly, there is some difficulty in relating evidence from the Venezuelan Andes and the Sierra Nevada de Santa Marta (Schubert 1972a, b, 1974; Gansser 1955; Raasveldt 1957) to the Sierra Nevada de Cocuy of Colombia (González et al. 1965). The suggested timing of the stage 4 moraines would put them with the Bolivariano of the Sierra Nevada de Santa Marta and the Mexican MV moraines (Gansser 1955; Raasveldt 1957; Heine 1975). From the assumed age, stage 3 moraines in the Sierra Nevada de Cocuy would fall into the general era of the Mexican M IV moraines. The stage 2 moraines of the Sierra Nevada de Cocuy might then correspond to Mamancanaca in the Sierra Nevada de Santa Marta (Gansser 1955; Raasveldt 1957) and M III in Mexico (Heine 1975). Stage 1 in the Sierra Nevada de Cocuy (González et al. 1965) is less readily compared. For the area of Nevado del Ruiz and Nevado del Tolima, Herd & Naeser (1974) bracket three moraine phases as younger than 13,760 B.P. These could thus correspond to the equivalents of the Mexican M III moraines (Heine 1975), but also M II and/or M IV. One outermost lateral moraine is dated as older than 100,000 B.P. (Herd & Naeser 1974).

A comparatively close spatial coherence of glacial-climatic events can be expected between the Ecuadorian Andes and the neighboring Cordillera Blanca of Peru. However absolute dating is also missing for that region, and the two proposed chronologies are conflicting (Clapperton 1972; Lliboutry et al. 1977). Table 5 presents the correlation between the two Cordillera Blanca classifications as arrived at from own field observations in Cordillera Blanca and literature evaluation. This differs from the scheme proposed by Lliboutry et al. (1977).

The moraines of complex I in Ecuador seem to correspond to the 'group 4' (Clapperton 1972) and 'Safuna VI' moraines (Lliboutry et al. 1977) of the Cordillera Blanca. The Ecuadorian II moraines could be the corollary to Clapperton's (1972) 'group 3' and Lliboutry et al.'s 'Safuna I/Huaraz m8'. For Lliboutry et al.'s 'Safuna IV-III/Huaraz m7' no equivalence is suggested with Clapperton's scheme for the Cordillera Blanca, nor with the moraine stages of Ecuador. Clapperton's 'group 2' appears to correspond to Lliboutry et al.'s 'Safuna II-I/Huaraz m5-m6'; and Clapperton's 'group 1' may be equivalent to Lliboutry et al.'s 'Safuna Huilcapampa&Huaraz m2-m3' and 'Safuna Upper Terrace/Huaraz m1-m4'. The latter two groups of Clapperton may in turn correspond to the complex III moraines and possible earlier glacial events in Ecuador.

Absolute dating for moraines in the Quelccaya/Vilcanota areas of Southeastern Peru (Mercer et al. 1975; Mercer & Palacios 1977) is interesting in relation to the Mexican M I-III and possibly IV moraines (Heine 1975), and the moraines 1-3 in Colombia (González et al. 1965; Herd & Naeser 1974). Correlation with and timing of the moraines in the spatially nearer regions is uncertain. Although Clapperton (1972) suggests correlations between the Cordillera Blanca and a region as distant as Patagonia, the correspondence with Northern Chile (Caviedes & Paskoff 1975) is not obvious.

For Eastern Africa (Table 4b) a plausible spatial correlation suggests itself between High Semyen, Ethiopia (Hastenrath 1974a, 1977b) and Mts. Kenya (Baker 1967) and Kilimanjaro (Downie & Wilkinson 1972), and the Ruwenzoris (de Heinzelin 1962; Livingstone 1962; Whittow et al. 1963; Osmaston 1965). Apart from one date on the deglaciation of the Ruwenzoris (Livingstone 1962), and the age determinations of the older glaciations of Kilimanjaro (Downie & Wilkinson 1972), absolute dates are lacking.

Personal field experience in both parts of the World suggests a correspondence between the 'younger moraines' at Mts. Kenya and Kilimanjaro with complex I in the High Andes of Ecuador; and of moraines VI A, B of Baker (1967) at Mt. Kenya, and moraines of the 'Recent Glaciation' (Downie & Wilkinson 1972) at Kilimanjaro, with moraines of complex II in Ecuador. It is noted, however, that two distinct phases are developed on both Mt Kenya and Kilimanjaro. The two groups of moraines of complex III in Ecuador may represent a corollary to moraines II-V at Mt. Kenya (Baker 1967) and the double sets of moraines of the 'Little Glaciation' at Kilimanjaro (Downie & Wilkinson 1972) – barring an absolute chronology. Conclusive evidence of older glacial events in the South American Andes is scarce (Herd & Naeser 1974). Sauer (1971) hypothesizes that tectonic activity played a major role for the glacial sequences in Ecuador.

For Australasia (Table 4c), absolute dating of moraines is surprisingly abundant – thanks to a small group of researchers primarily in Australia (Peterson & Hope 1972; Galloway et al. 1973; Hope et al. 1976; Loeffler 1972, 1976, 1979; Koopmans & Stauffer 1968). Spatial correlation offers a puzzling enough task within the region, and a matching with Eastern Africa and the Tropical Americas shall not be attempted here. A feature common to New Guinea (Peterson & Hope 1972; Galloway et al. 1973; Hope et al. 1976; Loeffler 1972, 1979) and East Africa (Livingstone 1962) is the deglaciation at some time before 12,600-14,000 B.P. A complete conformity of glacial-climatic events between the various parts of the tropics must not be expected. More abundant absolute dating is needed to clear up the spatial correspondence of glacial-climatic chronologies.

The present study may serve as a preliminary inventory of glaciation in the equatorial region of the New World. An early glaciation extended down to around 3500 m and lower, possibly at some time after 25,000 B.P., and retreated in several phases. This glacial event was followed by the deposition of thick layers of volcanic ash. Subsequently, glaciers advanced again down to around 4200 m. Ice retreat from these moraines may have begun a few 100 years ago, but around the middle of the 19th century at the latest. Recession, with indications of intermediate halts, has continued to the present. Earlier moraine stages are suggestive of corollaries in neighboring regions of the Tropical Americas, but confirmation hinges on absolute dating. Recent glacier behavior broadly parallels that in the two other high mountain regions under the Equator, East Africa and New Guinea.

8.5 RECOMMENDATIONS

Because of their extraordinary altitudinal range and their position bridging the two hemispheres, the Ecuadorian Andes appear especially attractive for research into (1) the pleistocene and early holocene glacial climatic history, and (2) the recent glacier behavior in low latitudes.

(1) The exploration of developments since pleistocene times should strive for the establishment of an absolute chronology by means of C14 and other dating techniques. In addition to attempts at sampling of datable material in the moraine regions themselves, a large-scale tephrostratigraphy based on the various volcanic ash layers offers promise for bracketing the age of major glacial events. Among the moraine regions which seem most suitable for the in situ sampling of datable organic matter are the East side of Chimborazo and the caldera sector of El Altar, in particular. Spatial correlation of glacial events with archaeological sites in the Quito basin and elsewhere through tephrostratigraphy is a further attractive prospect. More important perhaps is the relation of glacial episodes to vegetation history, which should be reconstructed by palynological coring at strategic locations. Van der Hammen's (1974) review for Northern South America opens a perspective for the Ecuadorian Andes. His paper is therefore reproduced in Appendix IV. Even though no unique correspondence must be expected between the glacier fluctuations in the peak regions and the variations of vegetation and of human settlement in the intramontane basins, a synopsis should be attempted of glacial chronology, palynology, and archaeology.

(2) The recent glacier behavior in low latitudes deserves particular attention,

in that mountain glaciers are extremely sensitive indicators of climate. Glacier variations are conspicuous, but the climatic factors which cause them are not generally known. In fact, the concomitant long-term variations in meteorological parameters may be too small to measure directly. Systematic glacier observations are therefore not only needed in assessing the regional hydrological budget, but they may also yield important insight into climatic change on a global scale.

The current knowledge of climate and glacier variations — and even of the present ice extent — is particularly deficient for the tropical belt. Recent efforts at the international level focus on the compilation of a World Glacier Inventory and the continuous assessment of glacier variations (UNESCO 1970; Temporary Technical Secretariat for World Glacier Inventory of UNESCO-UNEP-IUGG-IASH-ICSI 1977; International Association of Hydrological Sciences — UNESCO 1977). Appendix III summarizes information on Ecuadorian glaciers supplied for the World Glacier Inventory, in relation to Maps 1-16. Compilation of corresponding information is envisaged for all Andean countries and other parts of the tropics. Such inventories may provide a reference for specific field investigations of tropical glaciers.

The glaciers of Western New Guinea have been the object of the Australian Universities Expeditions (Hope et al. 1976; Allison & Kruss 1977). For the glaciers and ice fields of East Africa, a review has been completed of variations since the first observations in the latter part of the past century (Hastenrath 1975). In particular, a multi-annual field program is being conducted on Lewis Glacier, Mt. Kenya (Hastenrath 1975; Caukwell & Hastenrath 1977; Hastenrath & Caukwell 1979). More extensive glaciation is found in the tropical Andes.

A first inventory and mapping of glaciers in the Cordillera Blanca of Peru was accomplished by the small but singularly successful mountaineering and scientific expeditions of the Deutsch-Osterreichischer Alpenverein in the 1930's (Kinzl 1949). Awareness of environmental hazards resulting from avalanches and the breakout of moraine-dammed lakes has long prompted a systematic glacier monitoring program in the Cordillera Blanca of Peru (Morales Arnao 1969; Kinzl 1976). A multi-annual field program aimed at the retrieval of ice cores, assessment of the present climate, heat and mass budget, and reconstruction of a climatic history, is underway on the Quelccaya Ice Cap in Southeastern Peru (Mercer et al. 1975; Hastenrath 1978; Thompson et al. 1979).

For the Ecuadorian Andes, no glaciological observation program exists at present. Historical documentation of former ice conditions is exceptional for the tropics, in that it covers in varying detail the span of several centuries. A systematic monitoring of selected glaciers in the Ecuadorian Andes appears highly desirable.

The observation program should assess the present heat and mass budget, kinematics and morphology of glaciers, so as to permit a quantitative evaluation of long-term glacier variations in terms of the climatic forcing. In the selection of type glaciers the following properties appear desirable in particular: a well-defined catchment area, reasonable accessibility, and the existence of historical documentation on earlier glacier variations. A comprehensive observation program should be envisaged for a few glaciers which meet these criteria. The prime candidate is the Caldera Glacier (2) of El Altar. Some small glaciers on Carihuairazo may also be suitable. Furthermore, long-term changes in terminus position should be monitored for a large number of glaciers, the definition of the catchment area not being critical. In fact, it is desirable to obtain such data continuously for several glaciers originating from the same ice cap. The numerous glacier tongues of Cotopaxi are of interest here because of their easy accessibility from Quito. However, observations of selected glaciers at Chimborazo, Antisana, Cayambe, and other mountains are likewise desirable. The task could be accomplished with the cooperation of different agencies and individuals, and of the large mountaineering community, but with an established government institution assuming the responsibility for the overall project.

On the regional scale, such investigations would contribute to the assessment of the hydrological budget and provide quantitative insight into variations of climate. In the global perspective, these efforts would create important documentation on large-scale climate and glacier variations in the eminently interesting equatorial belt. It is hoped that the present study may provide the background for such endeavors.

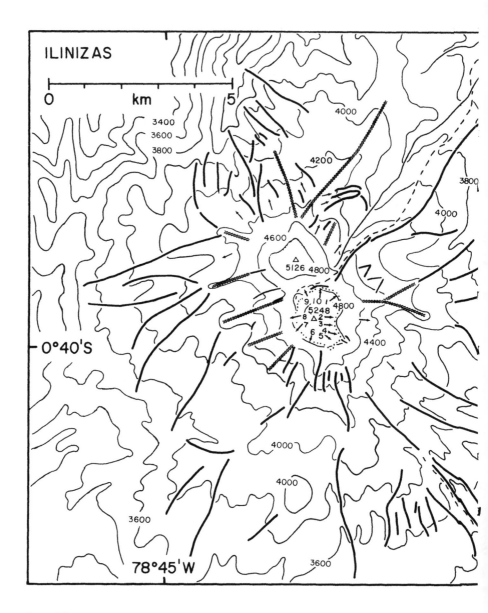

Legend for maps

1. Contour lines have 200 m spacing.
2. Moraines heavy solid lines.
3. Ice rim dotted line.
4. Travel route broken line.
5. Arrows indicate direction of present and/or former glaciers.
6. Present glaciers/ice lobes are numbered clockwise from around North.

00

3400

8°40' Map 2

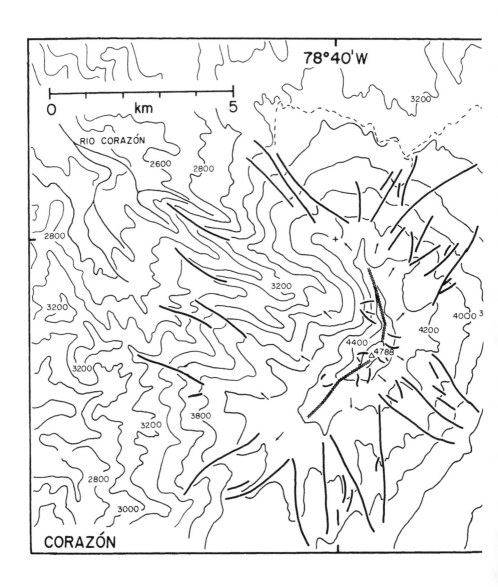

78°40'W

0 km 5

RIO CORAZÓN

2600 2800

2800

3200

3200

3200

3200

3800

3200

2800

3000

3200

4400

△4788

4200

4000 3

CORAZÓN

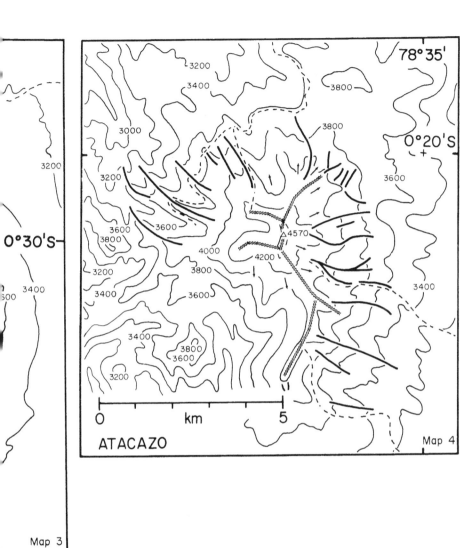

78°35'

3200
3400
3800
3000
3800
0°20'S
+
3200
3600
3600
3800
3600
4000
3200
4200
3800
3400
3600
3400
3400
3800
3600
3200

△4570

0 km 5

ATACAZO

Map 4

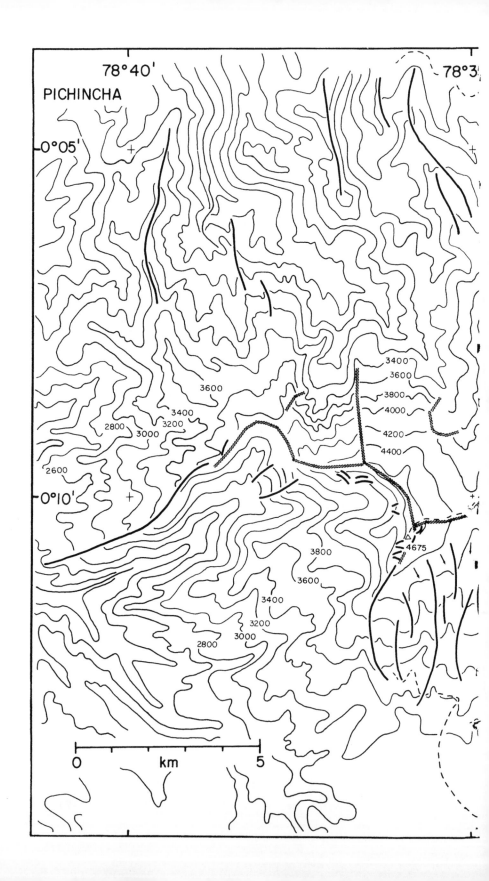

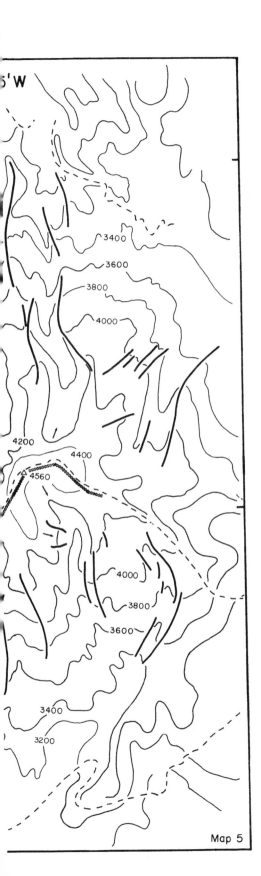

3400

3600

3800

4000

4200

4400

4560

4000

3800

3600

3400

3200

75

Map 5

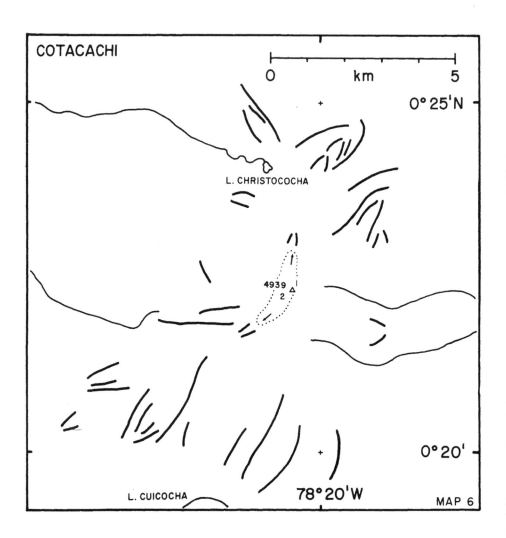

COTACACHI

0   km   5

0° 25'N

L. CHRISTOCOCHA

4939
2

0° 20'

L. CUICOCHA   78° 20'W

MAP 6

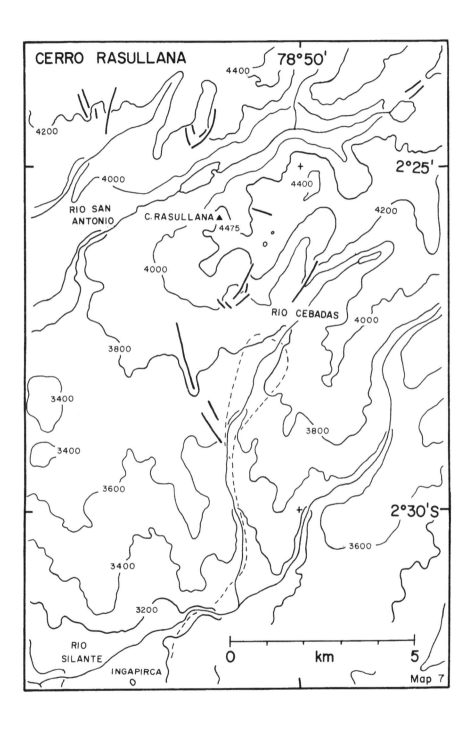

CERRO RASULLANA   78°50'

4400

2°25'

4200

4400

RIO SAN
ANTONIO   C. RASULLANA ▲
             4475

4000

4000

4200

RIO CEBADAS

4000

3800

3400

3800

3400

3600

2°30'S

3600

3400

3200

RIO
SILANTE

INGAPIRCA
○

0   km   5

Map 7

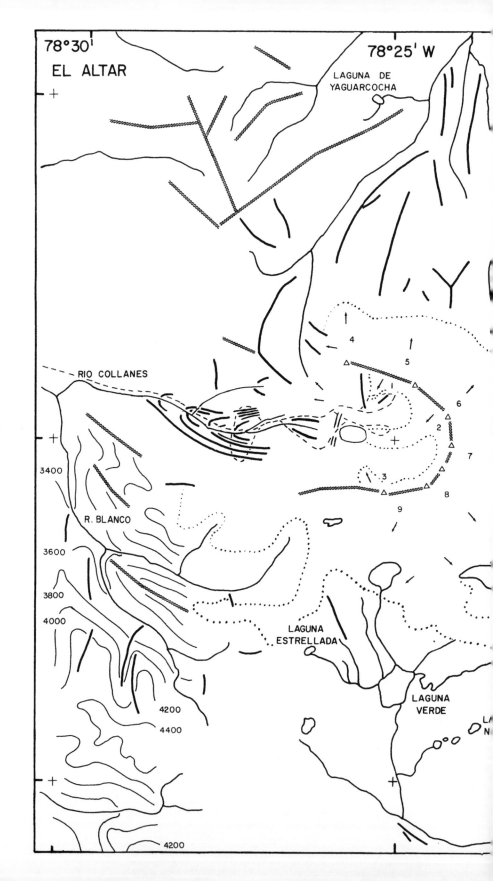

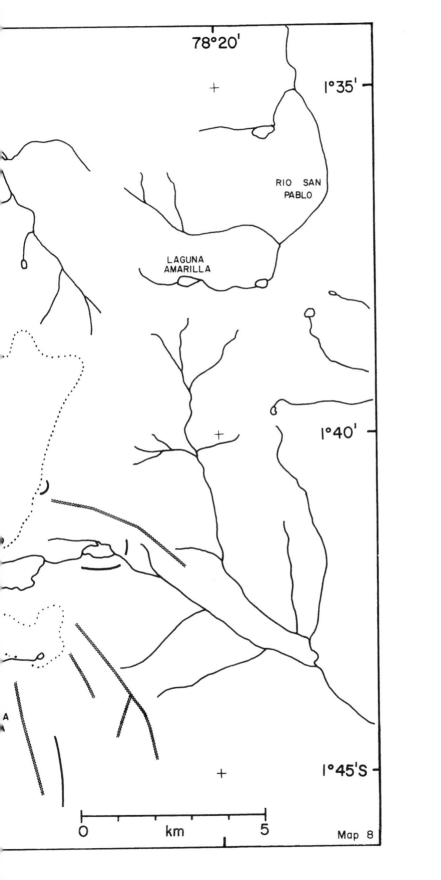

78°20'

1°35'

RIO SAN
PABLO

LAGUNA
AMARILLA

1°40'

1°45'S

0   km   5

Map 8

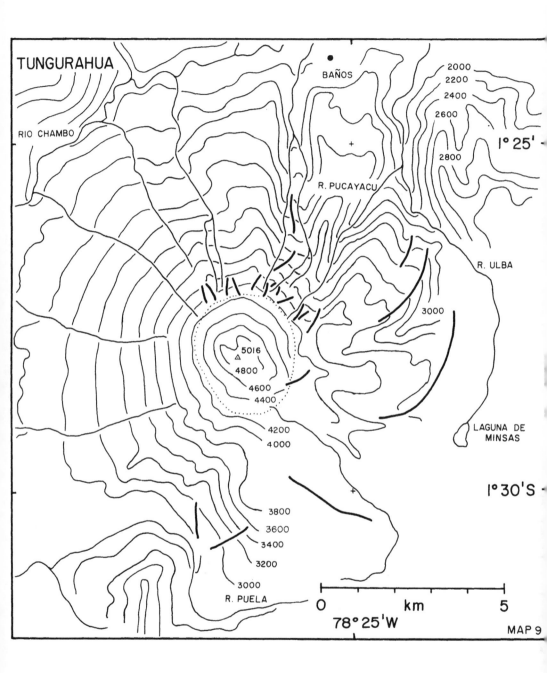

TUNGURAHUA

RIO CHAMBO

BAÑOS

2000
2200
2400
2600
2800

1° 25'

R. PUCAYACU

R. ULBA

3000

5016
△
4800
4600
4400

4200
4000

LAGUNA DE
MINSAS

1°30'S

3800
3600
3400
3200

3000
R. PUELA

0   km   5

78°25'W

MAP 9

80

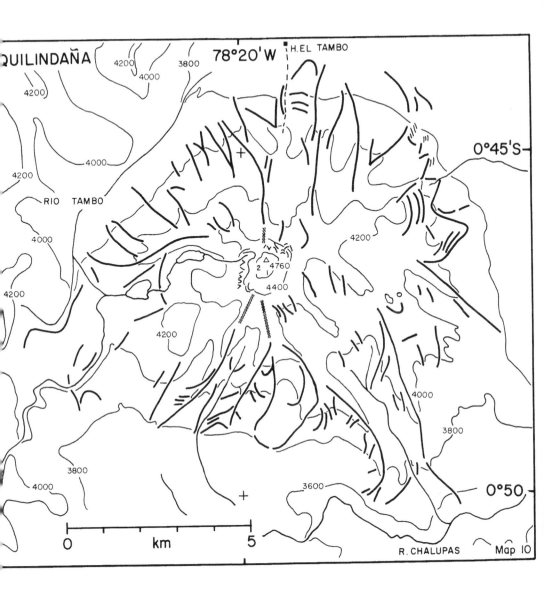

QUILINDAÑA 4200 3800 78°20'W ■H.EL TAMBO

4000

4200

0°45'S

4000

4200

RIO TAMBO

4000

4200

4200

△ 4760
2   4400

4200

4000

4200

3800

4000

3800

3600

4000   3800

3600   0°50

0   km   5

R. CHALUPAS   Map 10

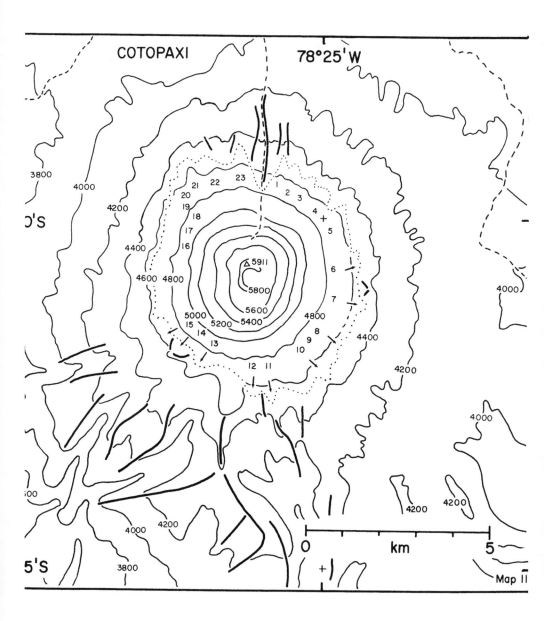

COTOPAXI   78°25'W

3800

4000

4200

0'S

4400

4600 4800

4800

5000

5200   5400

5600

5800

△ 5911

5911

21 22   23

20   1   2   3

19   4

18   5

17

16   6

                              7

21 22   23   1   2   3
                                  4
                                  5

                                  6

                                  7

                       4800
5000   5200   8
15   5400   9
14   10
13
        12 11

4000

4200

4000

4200

4000

4200   4200

4200

300

4000   4200

5'S   3800

0   km   5

+

Map II

82

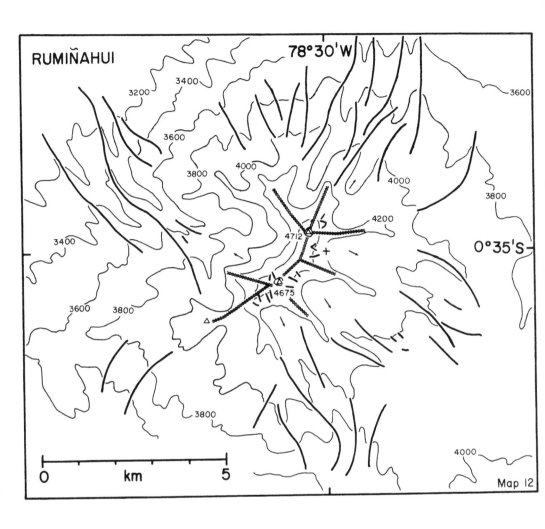

RUMIÑAHUI

78°30'W

0°35'S

3200
3400
3600
3600
3800
4000
4000
3800
4200
4712
3400
3600
3800
4675
3800
4000

0 km 5

Map 12

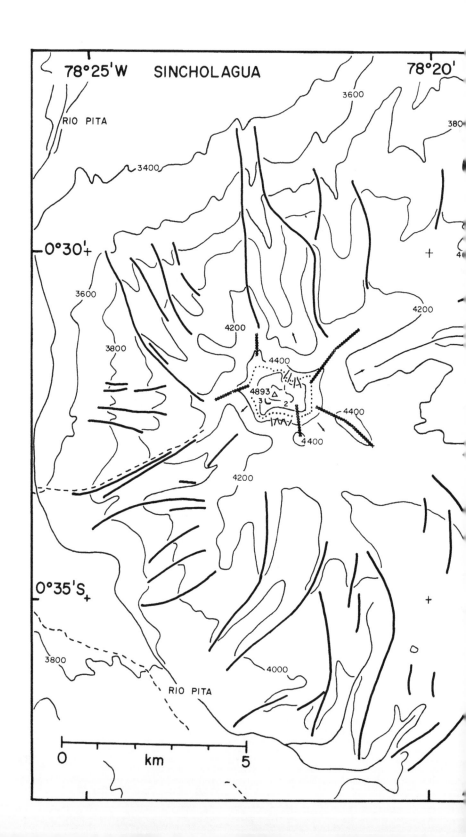

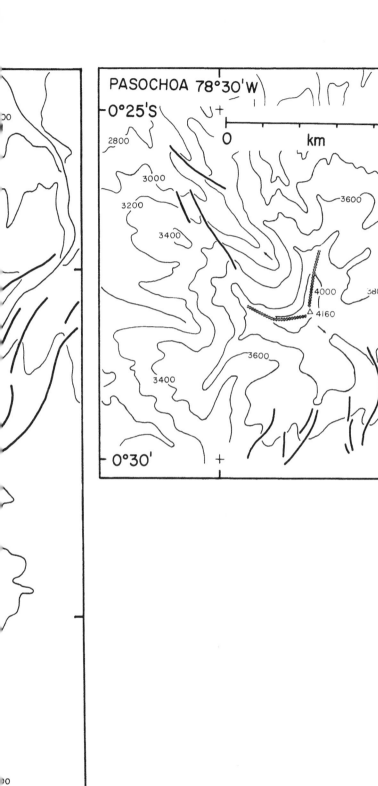

PASOCHOA 78°30'W

0°25'S

2800

3000

3200

3400

3400

3000

3400

3600

3400

4000

△ 4160

3800

3600

3400

0°30'

Map 14

Map 13

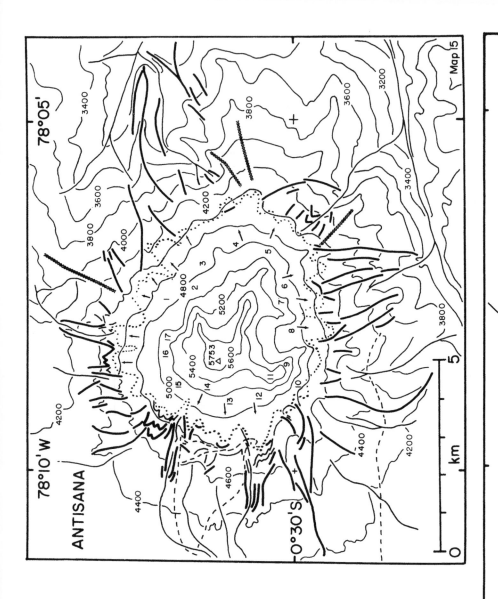

ANTISANA

CAYAMBE

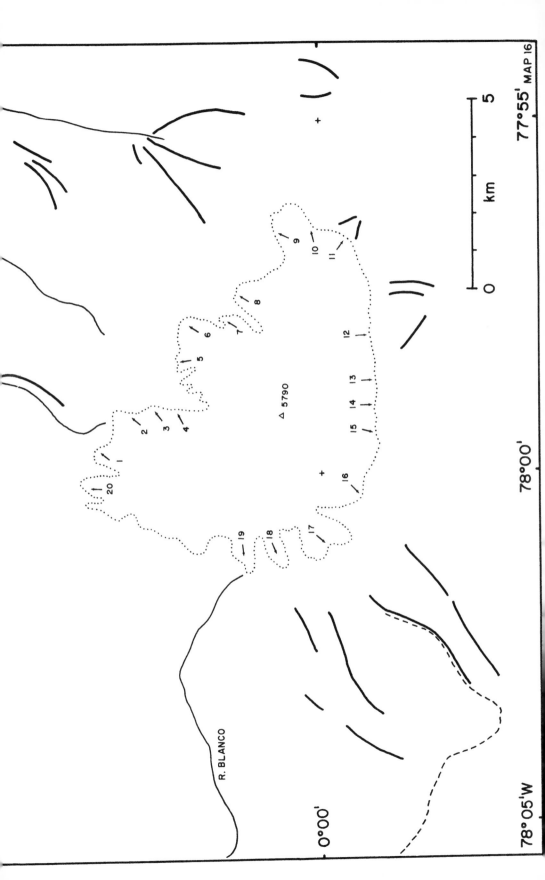

R. BLANCO

△ 5790

0°00'

78°05'W 78°00' 77°55' MAP 16

0 km 5

APPENDIX 1: TOPOGRAPHIC MAPS, AIR PHOTOGRAPHS, AND SATELLITE IMAGERY

A. WESTERN CORDILLERA

Maps

Air photographs

Chimborazo-Carihuairazo, *6 310 m,
                                *5 020 m

1:50,000:
Ñ IV-A3 (3890-III) (Simiatug)
Ñ IV-C1 (3889-IV) (Chimborazo)
Ñ IV-C3 (3889-III) (Guaranda)
Ñ IV-A4 (3890-II) (Ambato)
Ñ IV-C2 (3889-I)   (Quero)

USAF: VM, CAST-9, 1370 PMW,
AF 60-16, 12 Nov 63, rollo 71, #6313A-
6315A; VM, AST-2, 1370 PMW, 24 June 62,
rollo 27, #2048-2050; 24 June 62, rollo
26, #1878-1884; 26 June 62, rollo 26,
#1946-1949;

Ilinizas, *5126 m, *S 5248 m

1:50,000:
Ñ III-C3 (3892-III) (San Roque)
Ñ III-C4 (3892-II) (Machachi)
Ñ III-E1 (3891-IV) (Sigchos)
Ñ III-E2 (3891-I) (Mulaló)

USAF: VM AST-2 1370 PMW, 24 June 62,
rollo 26, #1855-1856;
USAF: VM CAST-9, 1370 PMW,
22 June 63, rollo 52, #4451-4453;

1:25,000 (not used as map base):
Ñ III-C3b (3892-III-NE) (Rio Zarapullo)
Ñ III-C3d (3892-III-SE) (Tungosillin)
Ñ III-C4c (3892-II-SW) (Iliniza)
Ñ III-E1b (3891-IV-NE) (Yaló)
Ñ III-E1d (3891-IV-SE) (Isinlivi)
Ñ III-E2a (3891-I-NW) (Pastocalle)

Corazón, 4788 m

1:50,000
Ñ III-C2 (3892-I) (Amaguaña)
Ñ III-C4 (3892-II) (Machachi)

USAF: VM CAST-9, 1370 PMW, AF 60-
16, 22 June 63, línea 39, rollo 52, #4447-
4448; VM 1370 PMW, AF 60-16, 18 Feb
66, línea 40, rollo 85, #7989-7992;

1:25,000 (not used as map base):
Ñ III-C2c (3892-I-SW) (El Pongo)

Ñ III-C4a (3892-II-NW) (Cerro Corazón)
Ñ III-C4c (3892-II-SW) (Iliniza)

Atacazo, 4570 m

1:50,000:
Ñ III-A4 (3893-II) (Quito)
Ñ III-C2 (3892-I) (Amaguaña)

1:25,000 (not used as map base):
Ñ III-C2a (3892-I-NW) (Rio Santa Ana)
Ñ III-C2b (3892-I-NE) (Amaguaña)

USAF: VM 1370 PMW, AF 60-16,
18 Feb 66, rollo 85, línea 40, #7986;
línea 41, #7936-7937;

Pichincha, 4675 m

1:50,000:
Ñ III-A2 (3891-I) (Nono)
Ñ III-A4 (3893-II) (Quito)

USAF: VM 1370 PMW, AF 60-16,
18 Feb 66, rollo 85, línea 41, #7941-7943,
7980-7981; línea 42, #8005-8006;

Cotacachi, *4939 m

1:50,000:
CC-Ñ III-D3 (Plaza Gutierrez), censo 40

1:25,000 (not used as map base):
plancheta 2 de hoja 28 (Cotacachi)

HYCON: 15 Feb 56, #124256-124257,
124268-124269;
USAF: 7 March 63, 2637-2638, 2653-2654;

Chiles, 4764 m

1:25,000:
plancheta 17 de hoja 13 (Chiles)

IGM JET: 19 Jan 78, línea 30, rollo 32,
#6538-6539;

EASTERN CORDILLERA

Maps

Air photographs

Rasullana, 4475 m

1:50,000:
Ñ-IV-C1 (3886-IV) (Juncal)
Ñ-V-C3 (3886-III) (Cañar)

USAF: VM, CAST-2, 1370 PMW, 24 June
62, rollo 27, línea 35, #2079-2082;
línea 35 A, #4866-4869;

Soroche, *4730 m

1:50,000:
Ñ-V-A4 (3887-II) (Totoras)
Ñ-V-C2 (3886-I) (Huangra)

1:100,000 geological map
sheet 71 (Alausi)

HYCON: 15 Feb 56, línea 543,
#29809-29812;
IGM JET; 4 Apr 77, línea 22-R-21,
#4107-4108; línea 23-R-21, #4028-4031;

Colay, *4630 m

1:50,000:
N-V-A4 (3887-II) (Totoras)

1:100,000 geological map
sheet 71 (Alausi)

HYCON: 15 Feb 56, línea 543,
#29809-29812;

IGM JET: 4 Apr 77, línea 22-R-21,
#4107-4108; línea 23-R-21, #4028-4031;

Sangay, *5230 m

USAF: VM, 1370, PMW, AF 60-16,
8 Feb 65, R-77, #6893-6895;

Cubillín, *4500 m, approximately

1:50,000:
Ñ IV-F1 (Huamboya), censo 187

USAF: 22 Apr 63, línea 44, #4688-4690;

IGM JET: 4 Apr 77, línea 23, rollo R-21,
#4041-4042;

Cotopaxi, *5911 m

1:50,000:
Ñ III-C4 (3892-II) (Machachi)
Ñ III-D3 (3992-III) (Sincholagua)
Ñ III-E2 (3891-I)   (Mulaló)
Ñ III-F1 (3991-IV) (Cotopaxi)

1:25,000 (not used as map base):
Ñ III-F1a (3991-IV-NW) (Cotopaxi)
Ñ III-D3c(3992-III-SW) (Rio Pita)

HYCON: VV, HY, M, 174 AMS, 25 Nov
56, 153, #29767-29768; VV, HY, M,
142 AMS, 15 Feb 56, 153, #124293-
124294, 124307-124309;

Rumiñahui, 4712 m

1:50,000:
Ñ III-C4 (3892-II) (Machachi)
Ñ III-D3 (3992-III) (Sincholagua)

1:25,000 (not used as map base):
Ñ III-C4b (3892-II-NE) (Machachi)
Ñ III-C4d (3892-II-SE) (Cotopaxi Minitrak)

HYCON: VV, HY, M, 174 AMS, 25 Nov
56, 153, #26764-26766;

Sincholagua, *4893 m

1:50,000:
Ñ III-D1 (3992-IV) (Pintag)
Ñ III-D3 (3992-III) (Sincholagua)

1:25,000 (not used as map base):
Ñ III-D1d (3992-IV-SE) (La Cocha)
Ñ III-D3d (3992-III-SE) (Lago Sinigcocha)
Ñ III-D3a (3992-III-NW)(Rayoloma)
Ñ III-D3b (3992-III-NE) (Sincholagua)

HYCON: VV, HY, M, 142 AMS, 15 Feb
56, 153, #124289-124292, 124312,
124234;

Altar, *5319 m

1:50,000:
Ñ IV-D3 (El Pungal), censo 180
Ñ IV-F1 (Huamboya), censo 187
partial contours

HYCON: 13 June 56, línea 541, #29595-
29599; línea 543, #29543-29545;
línea 543, #29792-29793.
IGM JET: 4 Apr 77, línea 24-R-21,
#4004-4011;

Tungurahua, *5016 m

1:50,000:
Ñ IV-D1 (Baños), censo 173,
partial contours
Ñ IV-D3 (El Pungal), censo 180,
partial contours

HYCON: 13 June 56, línea 541, #29602-
29604; línea 542, #29538-29540;
línea 543, #29788;

Cerro Hermoso, *4571 m

1:50,000:
Ñ-IV-B3 (3990-III) (Sucre)

HYCON: 13 June 56, línea 1-R-12,
#29652-29653;
IGM JET: 17 Sept 76, línea 1-R-12,
#2342-2344; línea 2-R-12, #2353-2355;

Quilindaña, 4760 m

1:50,000:
Ñ III-F1 (3991-IV) (Cotopaxi)
Ñ III-F3 (3991-III) (Laguna de Anteojos)

1:25,000 (not used as map base):
Ñ III-F4 (Chalupas), censo 152

USAF: VM, 1370 PMW, AF 60-16, 1 Feb
66, R 83 #7690-7691;
HYCON: VV, HY, 142 AMS, 15 Feb 56,
153, #124295-124297; #124227-124228;

Pasochoa, 4160 m

1:50,000:
Ñ III-C2 (3892-I)   (Amaguaña)
Ñ III-C4 (3892-II)   (Machachi)
Ñ III-D1 (3992-IV) (Pintag)
Ñ III-D3 (3992-III) (Sincholagua)

HYCON: VV, HY, M, 174 AMS, 25 Nov
56, 153, #29761-29763; VV, HY, M,
142 AMS, 15 Feb 56, 153, #124313-
124315;

Antisana, *5753 m

1:25,000:
Ñ III-D2-C (3992-I-SW) (Antisana)
Ñ III-D4-a (Laguna de Miracocha)
Ñ III-D2-d (Rio Quinjua)
Ñ III-D4-b (Rio Quijos)

1:50,000 (not used as map base):
Ñ III-D2 (Papallacta), censo 114
Ñ III-D4 (Laguna Miracocha), censo 130

HYCON: VV, HY, M, 142 AMS, 15 Feb
56, 153, línea 538, #1241256-124159;
línea 539, #124217-124220;
USAF: VM 1370 PMW, AF 60-16, 7 Feb
65, R-76, línea 49, #6730-6732; línea 52,
#6711-6714; 8 Feb 65, R-78, línea 52A
#7102-7104;

Sara Urcu, *4676 m

1:50,000:   USAF: 17 Nov 66, línea 56, #7601-7602;
CC 0 III-A1 (Sara Urcu), censo 80   IGM JET: 31 May 78, línea 30-D-R-33,
                                        #6766, 6782-6783;

Cayambe, *5790 m

1:50,000 m   USAF: 8 Feb 65, línea 52A, #7117-
CC Ñ II-F4 (Cayambe), censo 65   7119; 8 Feb 65, línea 54, #7177-7178;
CC O II-E3 (Laguna San Marcos),   17 Feb 66, línea 56, #7595-7598;
censo 66
CC Ñ III-B2 (Cangahua), censo 79
CC O III-A1 (Sara Urcu), censo 80

Satellite imagery (Landsat)

Image no.	Date	Hour (GMT)	MSS bands
10/61:2934-14213	13 Aug 1977	14:21	7 45 7 color
10/60:2934-14210	13 Aug 1977	14:21	7 45 7 color
10/59:21168-14182	4 Apr 1978	14:18	7 45 7 color
10/60:30345-14454	13 Feb 1979	14:45	7 45 7 color
10/61:30345-14460	13 Feb 1979	14:46	7 45 7 color

APPENDIX II: HISTORICAL SOURCES

GENERAL

1. Date 17 July 1535

González Rumazo, J. 1934a (*Cabildos de Quito*), Vol. 1, p. 128:
'mas se le señalo en XVII de julio de MDXXXV años bna estançia de puercos en el postrero ancon questa a la mano derecha de pançaleo ques el camyno que comyença en saliendo desta villa a man derecha del çerro gordo e a mano derecha de otro camyno que va a rriobanba junto a los syerros nevales.'

Schottelius, J.W. 1935-36 (*Die Gründung Quitos*), map:
This shows Pançaleo near Uyumbicho, a village at the foot of Atacazo.

2. Date 15 March 1540

González Rumazo, J. 1934a (*Cabildos de Quito*), Vol. 2, p. 105:
'En este dicho cabildo el dicho señor tenyente general. pidio a los dichos señores le conçedan e den vna estançia ques en termyno de esta villa entrel valle de chillo e pinta por la cabeçada de la estançia del tesorero rrodrigo nuñes e por la otra parte sobre la mano yzquyerda lynde con el rrio de pinta e por la otra parte lynde con vna quebrada grande e con estançia de (gregorio) diego ponze/hasya la sierra nebada/. frontero de quyxo/ . . . E luego el dicho señor tenyente general dixo que de mas de lo suso dicho pidio que por el rremate de la dicha estançia hasya la sierra nebada queda bn pedaço de tierra que no es para poder senbrarse de pan. ny ay (estan) otra estançia dada a ningund vezyno que le den la dicha añadidura ques hasta la sierra nevada e del anchor de lo demas/.'

Herrera y Tordesillas, A. de 1934 (*Historia*), Vol. 1, map at p. 123:
This shows 'Los Quixos' to the East of the Eastern Cordillera.

3. Date 1541

González Suárez, F. 1891 (*Historia del Ecuador*), Vol. 2, pp. 281-283:
'Gonzalo Pizarro . . . se puso en camino en los primeros meses del año de 1541 . . . El primer dia se detuvieron en un punto denominado Inga, que está á este lado de la cordillera oriental, y mientras no salieron de poblado el viaje fué cómodo y agradable; pero cuando principiaron a transmontar la gran cordillera, entonces comenzaron sus trabajos; muchos murieron, principalmente de los indios, helados de frío con el viento recio y húmedo de las alturas y la copiosa nevada que cayó mientras pasaban los expedicionarios. Al descender á la parte oriental al otro lado de la cordillera . . . llegaron á una población la primera de los Quijos, llamada Zumaco . . .'

4. Date 12 March 1550

González Rumazo, J. 1934b (*Cabildos de Quito*), Vol 4, p. 311:

'Este dicho dia en el dicho cabildo el dicho señor alcalde lorenço de çepeda dixo que pide a los dichos señores le fagan merçed de le mandar dar e prober dos estançias bna de bacas y otra de puercos/questan anbas en el termyno del pueblo de pinta que en el esta encomendado (van por la) que a por nonbre la dicha tierra que pide ychubanba y otra copal e otra clangli que lindan con tierras de diego rrodrigo e por la otra parte (tierra) la syerra (de) nebada de pinta y con los termynos del pueblo de pançaleo . . .'

5. Date 1738

La Condamine, C.M. de 1751 (*Journal*), p. 48:

'Pichincha & le Coraçon, sur le sommet desquels nous avons porté des baromètres, n'ont que 2430 & 2470 toises de hauteur absolue; & c'est la plus grande, que l'on sache, où l'on ait jamais monté. La neige permanente a rendu jusques ici les plus hauts somments inaccessibles. Depuis ce terme, qui est celui où la neige ne fond plus, même dans la zone torride, on ne voit guère, en descendant jusques a 100 ou 150 toises au dessous, que des rochers nuds, ou des sables arides; . . .'

6. Date 1727-1767

Velasco, J. de 1841-44 (*Historia*), Vol. 1, pp. 7-8:

'. . . los montes . . . pueden . . . dividirse en tres órdenes de altura. Los 18 de primer órden son todos perpetuamente cubiertos de nieve. De los del segundo se ven varios nevados por gran parte del año, y tal cual siempre. De los del tercero solo algunos y por poco tiempo. . .

Montes de 1.ª órden

Altar . . ., Cotacache . . ., Llanganate . . ., Mohanda (Otavalo), Pichincha, Purasé (Popayan), Rumiñahui, Saldaña (Quijos), Sangai, Sincholahua, Tungurahua, Yanaurco (Otavalo).

Montes de 2.ª órden

. . . Ashuai . . ., Carahuayrazo . . ., Collanes . . ., Corazon . . ., Saraurco

Montes de 3.ª órden

. . . Quelendana . . ., Quirotoa . . .'

Id. p. 10:

,Yanaurco: que quiere decir monte negro, tiene como quemado todo lo que no está cubierto de nieve, . . .'

7. Date 1802

Humboldt, A. de 1810 (*Vues des Cordillères*), p. 44:

'Le Cotopaxi est situé au sud-sud-est de la ville de Quito, à une distance de douze lieues, entre la montagne de Rumiñavi, dont la crête, hérissée de petit rochers isolés, se prolonge comme un mur d'une hauteur énorme, et le Quelendana, qui entre dans la limite des neiges éternelles.'

8. Date 1802

Humbodt, A. de 1853 (*Kleinere Schriften*), p. 21:

'Der Pichincha liegt . . . in derselben Achsenrichtung mit den Schneebergen Iliniza, Corazón und Cotocachi.'

9. Date 1802

Humboldt, A. de 1853 (*Kleinere Schriften*), p. 70:

'man sieht auf die mächtigen Schneeberge Cayambe, Cotocachi, Corazón, Iliniza, . . .'

10. Date 1802
   Humboldt, A. van 1874a (*Kosmos*), p. 381:
'Passuchoa, durch die Meierei el Tambillo vom Atacazo getrennt, erreicht so wenig als der letztere die Region des ewigen Schnees.'

11. Date 1802
   Villavicencio, M. 1858 (*Geografía*), p. 66:
'montañas con nieves perpetuas: Pichincha, Corazón, Atacazo; montañas con nieve ocasional o sin ella: Ruminagüi, Pasuchóa.'

12. Date 1858-59
   Wagner, M. 1870 (*Reisen*), pp. 627-628:
'Die untere Grenze des ewigen Schnees zeigt selbst in der Nähe des Aequators nach den verschiedenen Monaten und nach Lage und Form der Berge namhafte Abweichungen. Ich fand dieselbe am

Cotocachi im Mai	14,814 P.F.
Gagua-Pichincha im Juni	14,770
Mozo-Pichincha im Mai	14,791
Ilinissa im December	14,538
Carahuirazo im Januar	14,880
Tunguragua im Februar	14,650
Altar im Februar	14,876'

13. Date 1858-59
   Wagner, M. 1870 (*Reisen*), p. 628:

'Die Schneelinie ist am	nach Humboldt	nach Boussingault
Chimborazo	4816 Meter	4868 Meter
Cotopaxi	4853	4804
Antisana	4859	4871
Rucu (Mozo) Pichincha	4785	
Gagua Pichincha	4794	
Corazon	4790	

14. Date 1867
   Orton, J. 1870 (*Andes*), p. 125:
'Twenty-two summits are covered with perpetual snow . . . The snow limit at the equator is 15,800 feet.'
   Id., p. 142:
'Cotacachi is always snow-clad.'

15. Date 17-19 Aug 1870
   Dietzel, K.H. (ed.) 1921 (Reiss, *Reisebriefe*), p. 106:
'Der höchste Gipfel des Corazon ist ein kleines Plateau, das von einer kompakten, aber wenig mächtigen Schneemasse, aus der an vielen Stellen schwarzes Gestein herausragt, bedeckt ist. Einzelne Schneeflecke und Schneelehnen ziehen sich am Abhange herab. Zur Bestimmung der unteren Schneegrenze ist jedoch der Corazon ebensowenig wie der Pichincha geeignet.'

16. Date 1870-74
   Reiss, W. & Stübel, A. 1892-98 (*Hochgebirge*), Vol. 2, pp. 180-182:

'Höhen der Schneegrenze und der Gletscherenden an den Schneebergen Ecuadors nach
W.Reiss und A.Stübel 1871-74.

Name des Berges	Gipfelhöhe (Meter)	Schneegrenze	Gletscherenden
West-Cordillere:			
Cotacachi (XII. 70)	4966 t.R.		
Südwestseite		4705 b.R.	4597 b.R.
Ostseite		4694 t.	4537 t.
Ostseite		4620	
Südseite			4499
Rucu-Pichincha	4737		
Guagua-Pichincha	4787		
Corazon (VIII. 70)	4816	4679 b.	
Iliniza (XI. 72)	5305		
Nordwestseite		4771	
Westseite		4653	4484 b.
Carihuairazo (VII. 73)	5106		
Südseite		4675	
Ostseite			4386 b.St.
Ostseite			4354 b.R.
Nordseite			4500 b.R.St.
Chimborazo (VII. 73)	6310		
Nordseite		4862 b.R.	4255 b.R.(VI. 74)
Nordseite		4916	
Südwestseite			4358 b.R.
Südseite		4763 b.St.	
Südostseite		4714 b.R.	4550 b.R.
Südostseite			4516 b.St.
Ostseite		4616	4388
Ost-Cordillere:			
Cayambe (III. 71)	5840 t.R.		
Nordseite		4672 b.R.	4510 b.R.[1]
Nordseite			4400 St.
Nordostseite		4398	4134 R.
Ostseite			4298
Saraurcu (VII. 71)	4725 b.Whymper		
Westseite		4364 b.R.	4176 b.R.
Antisana (II. 72)	5756 t.R.		
Nordwestseite		4784 b.St.	
Nordseite		4721 R.	
Westseite		4694	
Südwestseite			4620 b.R.
Südwestseite			4618 St.
Südostseite			4216 R.
Sincholagua	4988 t.R.		
Nordseite		4577 b.St.	
Quilindana (IV. 72)	4919 t.R.		
Nordseite			4470 b.R.

98

Cotopaxi (IV. 72)	5943 t.R.		
Nordseite		4741 b.R.	
Nordwestseite		4763	
Westseite (XII. 72)		4627	
Südseite (XII. 72)		4629	
Ostseite		4626	4512 b.R.
Ostseite		4572	4300
Ostseite		4555	4230
Cerro hermoso (I. 73)			
Westseite	4576 t.R.		4242 t.R.
Tunguragua (III. 74)	5087 t.R.		
Nordwestseite		4600 b.St.	
Südseite			4272 b.St.
Südseite			4197 b.R.
Altar (IV. 74)	5404 t.R.		
Westseite			4028 b.St.
			3978 b.R.
Sangay (IX. 73)	5323 t.R.		
Südseite			4308 b.R.
Südostseite			4197 b.R.

t. = trigonometrische; b. = barometrische Messungen; R. = Reiss; St. = Stübel.
Das beigefügte Datum bezieht sich auf die Messungen von W. Reiss.
1. Unteres Ende der Endmoräne = 4305 m.

17. Date 1877
    Dressel, L. 1877 (*Vulkane*), p. 450:
'Die gipfel des Chimborazo, Cotopaxi, Cayambe, Antisana, Altar, Sangay, Iliniza, Carihuairazo, Tunguragua, Sinchulagua, Cotocachi, Quilindaña, Corazón, Azuay sind in ewigen, dichten Schnee gehüllt, während der Pichincha, Rumiñahui, Pasachoa, Atacatzo, Igualata, Imbabura, Yanaurcu und viele andere Berge, sowie ganze Gebirgs- rücken nur vorübergehend ein leichtes Schneegewand überwerfen.'

18. Date 1880
    Whymper, E. 1892 (*Travels*), p. 347:
'*Range south of Chimborazo* (15-16,000 feet). No permanent snow.
Chimborazo (20,498 feet). In January, little snow below 16,600 feet on the south side, but at that time it extended nearly one thousand feet lower on the E. and N. sides. In June-July there was deep snow as low as 15,600 feet on all these sides. At the same time, there was little snow below 16,700 feet upon the W. side.
Carihuairazo (16,515 feet). Very little snow below 15,000 feet in January, and much in June-July as low as 14,300 feet.
Corazon (15,871 feet). Much snow fell almost daily upon this mountain down to 14,500 feet, but there were no permanent snow-beds on the E. side, although there were some upon the W. side.
Atacatzo (14,892 feet). No permanent snow.
Pichincha (15,918 feet). The snow-beds were quite trifling in extent.
Cotocachi (16,301 feet). Permanent snow, in large beds, as low as 14,500 feet.
Imbabura (15,033 feet). Scarcely any snow below 16,000 feet on the west side. Covered with snow at 15,000 feet on the eastern side.
Sara-Urcu (15,502 feet). Snow fell daily upon this mountain lower than 14,000 feet, and was remaining permanently at about that elevation.

Antisana (19,335 feet). Permanently covered with snow as low as 15,300 feet.
Rumiñahui (15,607 feet). There was a small amount of permanent snow on the E., and none on the W. side.
Cotopaxi (19,613 feet). Snow fell frequently on Cotopaxi in February quite one thousand feet lower than it fell upon Chimborazo in January. It was remaining permanently on the western side at about 15,500 feet.
Llanganati group. Much snow below 16,000 feet.
Altar (17,730 feet). Many large snow-beds below 14,000 feet.
From examination of the above list, it will be seen that snow is in greater abundance upon the more easterly of the Great Andes.'

19. Date 1880
    Whymper, E. 1892 (*Travels*), pp. 348-349:
'Glaciers of large dimensions exist upon the Andes of the Equator. They attain their greates size upon Antisana, Cayambe, and Chimborazo, and there are considerable ones upon Altar, Carihuairazo, Cotocachi, Iliniza, Sara-urcu, and Sincholagua. There is also some very obscured glacier upon Cotopaxi. My glimpses of Quilindaña and Tunguragua were too slight to permit me to speak with certainty, but I believe that there are also glaciers upon those mountains.'

20. Date 1886
    Karsten, H. 1886 (*Géologie*), pp. 35-36:
'Les volcans dont les noms sont imprimés en caractères gras [= bold] dépassent la limite des neiges.

[bold print]		[medium print]	
Chiles	4780 m	Yanaurcu	4966 m
Corazon	4787	Rucu-Pichincha	4737
Iliniza	5305	Guagua-Pichincha	4787
Carihuairazo	5106	Atacatzo	4539
Chimborazo	6310	Pasechoa	4255
Cayambe-urcu	5840	Rumiñagui	4192
Antisana	5756		
Sincholagua	4988		
Cotopaxi	5943		
Qulilindaña	4919		
Tunguragua	5087		
El Altar	5404		
Sangay	5323		
Azuay	4600		

21. Date 1892
    Wolf, T. 1892 (*Geografía y geología*), p. 405:
'Si desatendemos algunos puntos aislados y muy reducidos sobre las crestas mas altas de la Cordillera oriental, en que se conserva la nieve durante todo el año, contamos en el Ecuador 16 cerros nevados:

En la Cordillera occidental:		En la Cordillera oriental:	
Chiles	4780 met.	Cayambe	5840 met.
Cotacachi	4966	Saraurcu	4725
Corazon	4816	Antisana	5756
Iliniza	5305	Sincholagua	4988
Carihuairazo	5106	Cotopaxi	5913
Chimborazo	6310	Quilindana	4919

Cerro hermoso	4576
Tunguragua	5087
Altar	5404
Sangay	5323

22. Date 1892
Wolf, T. 1892 (*Geografía y geología*), p. 406:
'. . . como en el Cerro hermoso y en el Saraurco. En el primero baja la línea de nieve à 4242 y en el segundo à 4364 metros. La línea de nieve más alta encontramos en el Chimborazo, cuyo clima es muy seco; se halla entre 4800 y 5000 m.'

23. Date 1903
Meyer, H. 1907 *(Hochanden)*, p. 289:
'Vom Rumiñagui und dem Pasochoa (4255 m) im Osten, vom Corazon (4787 m) und dem Atacatzo (4539 m) im Westen flankiert, hat der vom Rio grande durchströmte Südgipfel der grossen Quitomulde nur ungefähr 10 km Breite von Bergfuss zu Bergfuss . . . Von den genannten vier grossen Bergen trägt nur der Corazon (4787 m) etwas Schnee auf seinem hochgewölbten Felsgipfel . . . Da die Gipfelhöhe dicht an der bei 4800 m zu ziehenden mittleren Firngrenze dieser Landstriche liegt, so kann die Schnee-grenze nur auf orographisch begünstigte kleine Teile des Gipfels beschränkt sein.'

24. Date 1903
Meyer, H. 1907 (*Hochanden*), p. 428:
'Auf der Ostseite der Ostkordillere liegt (nach Reiss' Zusammenstellung) die Firngrenze bei 4480 m, auf ihrer Westseite bei 4660 m. Auf der Ostseite der Westkordillere liegt die Firngrenze bei 4670 m, auf ihrer Westseite bei 4710 m. . . . Seitdem die letzten dieser Messungen in den ecuatorianischen Anden gemacht worden sind, hat eine Ver-schiebung der Firngrenzen nach oben stattgefunden, die nach meinen Beobachtungen mindestens 50 m beträgt. Ich glaube daher die gegenwärtige mittlere Firngrenze auf etwa 4700 m für die Ostkordillere und auf etwa 4800 m für die Westkordillere nor-mieren zu können. Die mittlere Gletschergrenze liegt, wie nachher zu zeigen sein wird, noch etwa 300 m (nach Reiss vor 25 Jahren) bezw. 200 m (gegenwärtig) tiefer, also bei 4500 bis 4600 m. Die mittlere Grenze des Schneefalles aber kann bei 3700 m gezogen werden.'

CHIMBORAZO – CARIHUAIRAZO

1. Date 1858-59
Wagner, M. 1870 (*Reisen*), pp. 627-628:
See General 12.

2. Date 1858-59
Wagner, M. 1870 (*Reisen*), p. 628:
See General 13.

3. Date 22-24 July 1873
Dietzel, K.N. (ed.) 1921 (Reiss, *Reisebriefe*), p. 192:
(NW side of Chimborazo)
'Wir nähern uns ihm durch den Hondon de Llamacorral . . . Man kann hier die Höhe von 5,000 m erreichen, ohne in Schnee zu kommen . . .'

4. Date 1873
   Reiss, W. & Stübel, A. 1892-98 (*Hochgebirge*), Vol. 1, p. 234:
'Auf der Nordseite des Chimborazo liegt die Schneegrenze höher als an irgend einem
anderen Vulkanberge Ecuadors, denn auf den steilen Schutt- und Schlackenhalden von
Puca-huaico kann man an einigen Stellen bis über 5,000 m hoch aufsteigen, ohne
Schnee zu betreten.'

ILINIZAS

1. Date 1858-59
   Wagner, M. 1870 (*Reisen*), pp. 627-628:
See General 12.

2. Date 1870-74
   Reiss, W. & Stübel, A. 1892-98 (*Hochgebirge*), Vol. 2, pp. 180-182:
See General 16.

3. Date 1870-74
   Stübel, A. 1897 (*Vulkanberge*), pp. 62-63:

Südgipfel des Iliniza	5305 m
Nordgipfel des Iliniza	5162
Sattel zwischen beiden Gipfeln, Ostseite	4849
Rand des Sattels auf der Westseite	4600
Fuss des Gletschers, der den Sattel zwischen den beiden Gipfeln erfüllt, Westhang	4484
Untere Schneegrenze auf der Westseite bei Cucucuchu	4653
Untere Schneegrenze auf der Nordostseite des Südgipfels, Loma Milin	4771

4. Date 1870-74
   Reiss, W. & Stübel, A. 1892-98 (*Hochgebirge*), Vol. 2, pp. 169-170:
'Beide Gipfelpyramiden sind stark vergletschert, die südliche (5305 m) mehr als der
etwas niedrigere Nordgipfel (5162 m). Nach dem etwa 4800 Meter hohen Sattel
zwischen beiden Pyramiden ziehen Firnfelder und Gletscher herab, und vom Sattel
selbst erstreckt sich gegen Westen ein grosser Gletscher bis zu 4484 m, während die
Schneegrenze dort zu 4653 m gefunden wurde. Sowohl dieser wie auch alle anderen
Gletscher des Iliniza hängen steil am Abhang herab. Etwa 50 Meter unterhalb sind
Gletscherschliffe auf dem Gestein erhalten, und alte Moränen zeigen sich in einem
flachen Thal (cuchu), südlich von der Einsattelung.'

5. Date February 1880
   Whymper, E. 1892 (*Travels*), p. 133:
'Two glaciers have their origin on the upper part of the southern ridge of Iliniza. That
which goes westward, almost from its commencement, is prodigiously steep, and is
broken up into the cubical masses termed séracs. The other glacier, descending towards
the east, though steep, is less torrential.'

6. Date 1903
   Meyer, H. 1907 (*Hochanden*), pp. 282-284:
'Vom Rande des Firnmantels aber schiebt sich auf der Ostseite des Südgipfels ein
Gletscher vor, der das vor seiner stirn aus geweitete Kahr Cuchu-huasi nicht mehr

erreicht . . . Auf den Sattel zwischen beiden Gipfeln läuft von Südgipfel ein breiter kluftiger Gletscher aus . . . Dier Sattelgletscher züngelt auf der Ostseite in ein Hondon hinein . . . Einen einzigen Gletscher mit einem steilen Kahr darunter hat die Nordpyramide des Iliniza auf der Ostseite . . . aber auf dem Nordnordost- und Nordhang sind bei 4200 m mittlerer Höhe zwei Kahre eingetieft, in die von dem darüberliegenden Firnmantel her nur noch kurze Eiszungen herabhängen . . . Aber das grösste Kesseltal des ganzen Iliniza ist nach Reiss' und Stübels Beschreibung und nach Stübels Bildern das auf der Westseite unter dem Mittelsattel zwischen den beiden Schneegipfeln ausgetiefte Hondon de Cutucuchu. Von beiden Gipfeln strömen die dem Sattel zugewandten Eismassen nach diesem Hondon zusammen, dessen Boden eine mittlere Höhe von etwa 4250 m hat. Aber der Gletscher erreicht nur noch den Oberrand des Hondon, wo die Eisgrenze von Stübel zu 4484 m Höhe gemessen wurde; heute wird sie analog dem Rückgang aller übrigen Gletscher bei ca. 4600 liegen . . . Im Süden und Westen, unter dem höheren Südgipfel . . . liegt die Eisgrenze eitwas tiefer; im Norden und Osten, unter dem kleineren Nordgipfel liegt sie etwas höher . . . Auf der Osthälfte enden die kleinen, kurzen Hängegletscher durchschnittlich bei 4850 m. Durchweg ist starker Rückgang und viel frischer Moränenschutt zu beobachten.'

CORAZÓN

1. Date 17 July 1535
   González Rumazo, J. 1834a (*Cabildos de Quito*), Vol. 1, p. 128:
   See General 1.

2. Date 1738
   La Condamine, C.M. de 1751 (*Journal*), p. 48:
   See General 5.

3. Date 20 July 1738
   La Condamine, C.M. de 1751 (*Journal*), p. 58:
   'Le 20 nous allâmes fair l'expérience du baromètre beaucoup plus haut que l'endroit ou nous étions campés, c'est-à-dire, sur le Pic même du Coraçon, dont la pointe est toujours couverte de neige, & surpasse d'une quarantaine de toises le terme constant au dessus duquel la neige ne fond jamais.'

4. Date 1727-1767
   Velasco, J. de 1841-44 (*Historia*), Vol. 1, pp. 7-8:
   See General 6.

5. Date 1727-1767
   Velasco, J. de 1841-44 (*Historia*), Vol. 1, pp. 10-11:
   'Corazon . . . Rara vez se ve sin nieve, si bien la conserva perpetuamente en diversas partes de su mayor altura.'

6. Date 1802
   Humboldt, A. von 1853 (*Kleinere Schriften*), p. 21:
   See General 8.

7. Date 1802
   Humboldt, A. von 1853 (*Kleinere Schriften*), p. 70:
   See General 9.

8. Date 1802
   Humboldt, A. de 1810 (*Vues des Cordillères*), p. 273:
'La montagne du Corazón, couverte de neiges perpétuelles, a pris son nom de la forme de son sommet, qui est à peu près celle d'un coeur.' (also Planche 51).

9. Date 1858
   Villavicencio, M. 1858 (*Geografía*), p. 66:
See General 11.

10. Date 1858-59
    Wagner, M. 1870 (*Reisen*), p. 628:
See General 13.

11. Date 1867
    Orton, J. 1870 (*Andes*), pp. 125, 142:
See General 14.

12. Date 17-19 August 1870
    Dietzel, K.N. (ed.) 1921 (Reiss, *Reisebriefe*), p. 106:
See General 15.

13. Date 1870-74
    Stübel, A. 1897 (*Vulkanberge*), p. 55:
'Seine Vergletscherung macht sich aus der Entfernung so wenig bemerkbar, dass man den Corazón kaum noch zu den Schneebergen Ecuadors zählen kann.

Gipfel des Corazón	4787 m
Grenze des Firnschnees auf der Nordseite	4697 m'

14. Date 1877
    Dressel, L. 1877 (*Vulkane*), p. 450:
See General 17.

15. Date 1880
    Whymper, E. 1892 (*Travels*), p. 110:
'Though the summit was free from snow, and there was none on the eastern side, there was much in gullies on the western side, and we fancied there might be considerable beds or even a glacier below. Viewed from the west this mountain would be considered to be within the snow-line.'

16. Date 1880
    Whymper, E. 1892 (*Travels*), p.347:
See General 18.

17. Date 1886
    Karsten, H. 1887 (*Géologie*), pp. 35-36:
See General 20.

18. Date 1892
    Wolf, T. 1892 (*Geografía y geología*), p. 84:
'Es un cerro hermoso con la cúspide cubierta de nieve perpétua.'

104

19. Date 1892
Wolf, T. 1892 (*Geografía y geología*), p. 405:
See General 20.

ATACAZO

1. Date 17 July 1535
González Rumazo, J. 1834a (*Cabildos de Quito*), Vol. 1, p. 128:
See General 1.

2. Date 15 March 1540
González Rumazo, J. 1834a (*Cabildos de Quito*), Vol. 2, p. 105:
See General 2.

3. Date 1541
González Suarez, F. 1891 (*Ecuador*), Vol. 2, pp. 281-283:
See General 3.

4. Date 12 March 1550
González Rumazo, J. 1834b (*Cabildos de Quito*), Vol. 4, p. 311:
See General 4.

5. Date 1802
Humboldt, A. von 1874a (*Kosmos*), p. 381:
See General 10.

6. Date 1858
Villavicencio, M. 1858 (*Geografía*), p. 66:
See General 11.

7. Date 1877
Dressel, L. 1877 (*Vulkane*), p. 450:
See General 17.

8. Date 1886
Karsten, H. 1886 (*Géologie*), pp. 35-36:
See General 20.

9. Date 1892
Wolf, T. 1892 (*Geografía y geología*), p. 84:
'Sigue el Atacazo, que con 4539 metros de altura no alcanza la región de la nieve perpétua.'

10. Date 1903
Meyer, H. 1907 (*Hochanden*), p. 289:
See General 22.

PICHINCHA

1. Date 17 July 1535
González Rumazo, J. 1934a (*Cabildos de Quito*), Vol. 1, p. 128:

Schottelius, J.W. 1935-36 (*Die Gründung Quitos*), map:
See General 1.

2. Date 1570-74
   Rodriguez de Aguayo, P. 1965 (*Quito*), Vol. 2, p. 202 (see also Vol. 1, pp. 46-47):
   'Tiene a la redonda de sí la dicha ciudad de Quito algunos cerros muy altos y redondos
   a manera de montón de trigo, de los cuales algunos dellos están todo el año cubiertos
   de nieve y echan humo noche y día y algunas veces llamas de fuego grandes; especial-
   mente el que está a las espaldas de la dicha ciudad de Quito, hacia los Yumbos, tres
   leguas de la dicha ciudad, . . .'

3. Date July 1738
   La Condamine, C.M. de 1751 (*Journal*), p. 48:
   See General 1.

4. Date 1738
   Bouguer, P. 1749 (*La figure de la terre*), pp. 14-15:
   'La hauteur du sommet pierreux de Pichincha est à peu près celle du terme inférieur
   constant de la neige dans toute les montagnes de la Zone torride.'

5. Date 1738
   Juan, J. & Ulloa, A. de 1748 (*Relación historica*), p. 306:
   '. . . el cual era lo más encumbrado de un Cerro de peña, que se levantaba quasi 200.
   Tuessas sobre lo mas alto del Páramo de Pichincha, que formando en su eminencia
   diferentes Puntas, ò Picachos, era el de mayor elevacion, el que entonces teniamos por
   Morada; todo èl cubierto continuamente de Yelo, y Nieve, y no menos vestida de uno,
   y otro nuestra Choza.'

6. Date June 1742
   La Condamine, C.M. de 1751 (*Journal*), p. 152:
   'Un grand nombre d'Indiens de Quito, qui vont tous les matins chercher à Pitchincha
   de la neige pour apporter a la ville, avoient passé tout proche d'eux.'

7. Date June 1742
   La Condamine, C.M. de 1751 (*Journal*), p. 155:
   (crater of Pichincha)
   'La neige n'étoit pas fondue par-tout, elle subsistoit dans quelques endroits; mais les
   matières calcinés qui s'y mêloient, & peut-etre les exhalaisons du volcan, lui donnoient
   une couleur jaunâtre:'

8. Date 1727-1767
   Velasco, J. de 1841-44 (*Historia*), Vol. 1, pp. 7-8:
   See General 6.

9. Date 1802
   Humboldt, A. von 1853 (*Kleinere Schriften*), p. 64:
   'Rucu-Pichincha reicht kaum 35 Toisen hoch über die ewige Schneegrenze hinaus, und
   einige Male habe ich ihn von Chillo aus völlig schneefrei gesehen.'

10. Date 1802
    Humboldt, A. von 1874a (*Kosmos*), pp. 189-190:

'Der eigentliche Feuerberg (Vulkan) wird der Vater oder Alte, Rucu-Pichincha, genannt. Er ist der einzige Theil des langen Bergrückens, welcher in die ewige Schneeregion reicht: also sich zu einer Höhe erhebt, welche die Kuppe von Guagua-Pichincha, dem Kinde, etwa um 180 Fuss übersteigt.'

11. Date 1802
   Humboldt, A. von 1810 (*Vues des Cordillères*), p. 291:
'On distingue dans mon dessin (1), Rucupichincha ou les sommets qui entourent le cratère; . . . la cime rocheuse de Guaguapichincha (4);' (also Planche 41)

12. Date 1858
   Villavicencio, M. 1858 (*Geografía*), pp. 64-65:
'En el siglo anterior los académicos franceses que se ocupaban en operaciones geodésicas en Quito, se quejaban de la nieve que los cubria en la estacion del guagua Pichincha, punto de donde hace muchos años que desapareció la nieve.'

13. Date 1858
   Villavicencio, M. 1858 (*Geografía*), p. 66:
See General 11.

14. Date 1858-59
   Wagner, M. 1870 (*Reisen*), pp. 627-628:
See General 12.

15. Date 1858-59
   Wagner, M. 1870 (*Reisen*), p. 628:
See General 13.

16. Date 1859
   Jameson, W. 1861 (*Journey*), pp. 184-185:
'Snow frequently falls on the sandy desert of the crater; but two or three days of fine weather cause its disappearance, excepting in some localities where it lies in patches, sheltered from the rays of the vertical sun. The summit of Pechincha barely enters the snow-limit . . .'

17. Date 17-19 August 1870
   Dietzel, K.N. (ed.) 1921 (Reiss, *Reisebriefe*), p. 106:
See General 15.

18. Date 1870-74
   Reiss, W. & Stübel, A. 1892-98 (*Hochgebirge*), Vol. 2, pp. 180-182:
See General 16.

19. Date 1877
   Dressel, L. 1877 (*Vulkane*), p. 450:
See General 17.

20. Date 1880
   Whymper, E. 1892 (*Travels*), p. 347:
See General 18.

21. Date 1903
Meyer, H. 1907 (*Hochanden*), p. 305:
'Auf den beide Hauptgipfeln des Pichincha, dem Guagua-Pichincha (4787 m) und dem Rucu-Pichincha (4737 m) fällt zwar oft und mitunter viel Schnee, aber er überdauert die trocknen Monate nur an wenigen orographisch begünstigten Stellen.'

COTACACHI

1. Date 1870-74
Reiss, W. & Stübel, A. 1892-98 (*Hochgebirge*), Vol. 2, pp. 180-182:
See General 16.

2. Date 1870-74
Stübel, A. 1897 (*Vulkanberge*), p. 89:

Gipfel des Cotacachi, Nordwestspitze	4966 m
untere Schneegrenze auf der Ostseite	4694
untere Schneegrenze auf der Südwestseite	4620
Ende des Gletschers auf der Ostseite	4537
Ende des Gletschers von Tiucungo	4597
Ende des Gletschers auf der Südseite	4499

3. Date April 1880
Whymper, E. 1892 (*Travels*), p. 261:
'. . . and has a face on the east (facing the basin of Imbabura) that is precipitous; and another less abrupt one on the west, largely covered with snow.'
    Id., p. 263:
'The true summit of Cotacachi is a pointed peak of lava, broken up by frost, extremely steep at the finish, and upon that account bearing little snow.'
    Id., p. 264:
'. . . it is not unlikely that a crater lies buried beneath the glacier which at present occupies the depression between its two peaks.'

SANGAY

1. Date 1727-1767
Velasco, J. de 1841-44 (*Historia*), Vol. 1, p. 9:
'Sangai . . . Tiene la boca por la parte meridional, casi descubierta de nieve, y sus erupciones no hacen daño a los poblados.'

2. Date 1857
Spruce, R. 1961 (*Llanganati*), p. 163:
'To the . . . South rose Sangay, . . . remarkable for its exact conical outline, for the snow lying on it in longitudinal stripes (apparently of no great thickness) . . .'

3. Date 1870-74
Reiss, W. & Stübel, A. 1892-98 (*Hochgebirge*), Vol. 2, pp. 180-182:
See General 14.

4. Date January 1880
   Whymper, E. 1892 (*Travels*), p. 73:
   'There were large snow-beds near its summit, but the apex of the cone was black, and was doubtless covered with fine volcanic ash.'

5. Date July 1929
   Moore, R.T. 1930 (*Ascent*), p. 228:
   '. . . the last half mile of altitude is covered with a crown of ice and snow . . .'

EL ALTAR

1. Date 1870-74
   Stübel, A. 1897 (*Vulkanberge*), p. 237:
   'An die weite thorartige Oeffnung des Kraterkessels schliesst sich das Valle de Collanes an, . . . Die untere Grenze der permanenten Eisbedeckung geht also hier bis zu 4300 m hinab; das Gletscherende − allerdings bedingt durch gute Bodenverhältnisse − sogar bis zu 4000 Metern.'

2. Date 1870-74
   Reiss, W. & Stübel, A. 1892-98 (*Hochgebirge*), Vol. 2, p. 170:
   'Aus dem . . . Krater des Altar quillt . . . ein gewaltiger Gletscher hervor, dessen unteres Ende im flachen Thalgrund von Pasuasu oder Collanes bis zu 4000 Meter absoluter Höhe herabreicht. Der Gletscher wird gespeist durch viele an den Innenwänden des Kraters herabhängende Gletscher und Firnfelder, und ähnliche Gletscher bedecken die Aussengehänge der Kraterumwallung.'

3. Date 1872
   Stübel, A. 1897 (*Vulkanberge*), p. 237:
   'Gerade diese Stufe, an welcher sich der Gletscher aufgebrochen und cascadenartig herabgestürzt zeigt, macht das Bild so überaus grossartig und malerisch. Die Eiscascade wird zu beiden Seiten von moränenartigen Schutthalten begrenzt. An der Bruchfläche des Gletschers lässt sich seine Mächtigkeit schätzen: sie beträgt 60-100 Meter.'

4. Date May 1874
   Dietzel, K.N. (ed.) 1921 (Reiss, *Reisebriefe*), p. 140:
   'Von allen Seiten stürzen fortwährend . . . Schneemassen in ihn hinab und häufen sich dort zu einem ausgedehnten Firnfelde an, das einen langen, mächtigen Gletscher entsendet, der gegen Westen zieht und dort über eine ca. 1500 Fuss hohe, fast senkrechte Wand, die den Kraterboden (Plazapamba, 4330 m) von dem Sumpftal Collanes trennt, hinabfällt. Ein Teil der Masse ist dabei abgerissen und liegt in Trümmern am Fuss der Felsen bei dem 'Pasuasu' genannten Wäldchen; ein anderer Teil senkt sich unzerbrochen hinab und erreicht als zusammenhängende Eismasse den Grund. Es ist dies der am weitesten herabgelangte Gletscher Ecuadors (4028 m) . . . Der Gletscher, der sich, in nördlichen Teile abgebrochen, im südlichen zusammenhängend, am oberen Talschluss über die Stufe herabsenkt, ist von zwei mächtigen Schutthalden eingefasst, die in weitem Bogen seine Stirn umgeben und sich als schmale Rücken aus dem Krater zum Gletscherfuss herabziehen. Sie stehen weiter auseinander, als es die heutige Breite des Eises bedingen würde, der Gletscher scheint also früher mächtiger gewesen zu sein.'

5. Date 17 June 1880

Whymper, E. 1892 (*Travels*), p. 305:

,The walls of the cirque are exceedingly rugged, with much snow, and the floor is occupied by a glacier, which is largely fed by falls from 'hanging-glaciers' on the surrounding slopes and cliffs.'

6. Date 19 June 1880

Whymper, E. 1892 (*Travels*), p. 307:

'The face towards the north carried several hanging-glaciers. Frequently heard the roars of avalanches tumbling from them on to the glacier in the crater, the true bottom of which probably lies several hundred feet below the ice. This crater-glacier, in advancing, falls over a steep wall of rock at the head of the Valley of Collanes, in a manner somewhat similar to the Tschingel Glacier in the Gasteren Thal. Some of the ice breaks away in slices, and is re-compacted at the base of the cliff, while part maintains the continuity of the upper plateau with the fallen and smashed fragments. This connecting link of glacier (seen in front) appears to descend almost vertically.'

7. Date 1903

Meyer, H. 1907 (*Hochanden*), pp. 173-174:

'Als 30 Jahre vor mir die Herren Reiss und Stübel hier weilten . . . reichte der Kratergletscher noch in einer imposanten Eiskaskade bis an den Fuss der Felsstufe zwischen die beiden Moränenwälle herab (4028 m nach Stübel; 3978 m nach Reiss), wo die abgestürzten Eismassen einen kleinen regenerierten Gletscher bildeten. Auf einem der von Stübels Begleiter Troya gemalten Ölbilder (siehe Bilderatlas Tafel 18) ist dieser Zustand eindrucksvoll dargestellt. Auch Whymper sah noch 1880 den Kratergletscher in einer Eiskaskade . . . über die Frontfelswände fallen und an deren Fuss in einen regenerierten Gletscher übergehen, der durch die Eiskaskade immer noch mit dem obern Gletscher in direktem Zusammenhang war. Jetzt endet der Gletscher 300 m höher oben am Oberrand der Felsstufe; über diese stürzt kein Eis mehr herab, und der regenerierte Gletscher am Fuss der Eiswand ist verschwunden. Nur eine flache Halde von frischem Moränenschutt verrät seine einstige Stätte.'

8. Date 1903

Meyer, H. 1907 (*Hochanden*), p. 180:

'Reiss und Stübel haben 1872 die Kicke des Gletschers am Rande der Pasuasufelswand, über die er damals noch ins Collanestal abstürzte, auf 60-100 m gemessen, Jetzt ist er an seinem Ende nur noch ca. 20 m dick. Die Stirn ist ganz flach und erhebt sich nirgends mehr über den Rand der Pasuasuwand. Wie hoch er in seinen mittleren Teilen den Caldraboden bedeckt, ist nicht genau zu schätzen; nach dem Neigungswinkel des Eisstroms vom Anfang bis zur Stirn dürfte er dort kaum noch 50 m dick sein, während Whymper 1880 den Calderaboden noch 'einige Hundert Fuss' dick unter der Gletscheroberfläche annahm. Der Körper des Gletschers ist jetzt eingesunken und eingebrochen, was allein schon neben seiner kolossalen Schuttbedeckung erkennen lassen würde, dass er stark im Schwinden begriffen ist.'

TUNGURAHUA

1. Date 1727-1767

Velasco, J. de 1841-44 (*Historia*), Vol. 1, pp. 7-8:

See General 6.

2. Date 1727-1767
   Velasco, J. de 1841-44 (*Historia*), Vol. 1, pp. 9-10:
   'Tungurahua . . . no obstante ser casi derecho, como una pirámide, todo lo que no está
   cubierto de nieve, lo tiene cubierto de elevado bosque.'

3. Date 18 April 1873
   Dietzel, K.N. (ed.) 1921 (Reiss, *Reisebriefe*), p. 199:
   'Ein Asyl in solcher Höhe – unser Zelt lag 4498 m über dem Meere – war uns nichts
   Ungewohntes . . . Zunächst mussten wir noch etwa 150 m über Geröllschutt aufsteigen,
   ehe wir den Schnee ereichten. Er lag an der Stelle, wo wir ihn betraten nur 1-2 m hoch,
   bestand aber aus deutlich unterscheidbaren Schichten verschiedenen Alters.'

4. Date 1870-74
   Stübel, A. 1897 (*Vulkanberge*), p. 254:
   'Er ist in seinem höheren Theile stark vergletschert, in seinem tieferen, besonders auf
   der Nordseite, nur mit geschichteten Schneemassen überdeckt. Auf der Südseite reicht
   die Schneebedeckung 300 Meter tiefer hinab, als auf der Nordseite, wo sie unter
   gewöhnlichen Witterungsverhältnissen nicht über 300 Meter Ausdehnung nach oben
   besitzt.'

CERRO HERMOSO

1. Date 1857
   Spruce, R. 1861 (*Llanganati*), p. 168:
   'Gran Volcan del Topo, or Yurag-Llanganati: It is the only one of the group which rises
   to perpetual snow, though there are many others rarely clear of snow. . .'

2. Date 8 July 1873
   Dietzel, K. N. (ed.) 1921 (Reiss, *Reisebriefe*), p. 182 (also Reiss, 1875, p. 287):
   'Wenn man den Cerro hermoso nur von der Westseite betrachtet, so begreift man nicht,
   wie sich auf ihm ein Gletscher bilden kann. Er nährt sich jedoch aus den grossen Firn-
   massen, die sich auf einem etwas gegen Süden geneigten Plateau anhäufen, denn seine
   Gipfelfläche dehnt sich, wie man deutlich von einem mehr südlich gelegenen Punkte,
   zum Beispiel von Mocha aus, sehen kann, von Westen nach Osten. Schon Dr. Stübel hob
   die interessante Tatsache hervor, dass die Schneegrenze in der Kordillere nach Osten zu
   sich immer tiefer herabsenkt. Der Cerro hermoso ist deshalb, obwohl er sich nicht bis
   zu 4600 m, also zur Höhe der unteren Schneegrenze in der Westkordillere, erhebt,
   dennoch nicht nur mit ewigem Schnee bedeckt, sondern hat sogar echte Gletscher mit
   Firnfeldern und blauem, kompaktem Gletschereis.'

3. Date 1870-74
   Reiss, W. & Stübel, A. 1892-98 (*Hochgebirge*), Vol. 2, p. 88:
   '. . . fand ich . . . an dem noch weiter ostwärts sich erhebenden Cerro hermoso de los
   Llanganates das Gletscher-Ende in 4242 m.'

4. Date 1870-74
   Stübel, A. 1897 (*Vulkanberge*), p. 144:
   'Aus dem Gebiet der Cerros de los Llanganates . . . haben wir nur den zackigen, schnee-
   bedeckten Cerro Hermoso . . . in einer kleinen Nebenskizze charakterisirt.'

5. Date 1892
   Wolf, T. 1892 (*Geografía y geología*), p. 405:
   See General 21.

6. Date 1892
   Wolf, T. 1892 (*Geografía y geología*), p. 406:
   See General 22.

7. Date 1933-34
   Andrade Marín, L. 1936 (*Llanganati*), p. 60:
   '. . . acerca del Cerro Hermoso o Yurac-Llanganati, creo que puedo afirmar . . . que no
   es un 'nevado perpetuo' . . .'

8. Date 1971
   Kennerley, J.B. & Bromley, R.J. 1971 (*Llanganati*), p. 3:
   'The permanent snow line is at about 4350 metres on Cerro Hermoso.'

QUILINDAÑA

1. Date 1727-1767
   Velasco, J. de 1841-44 (*Historia*), Vol. 1, pp. 7-8:
   See General 6.

2. Date 1802
   Humboldt, A. de 1810 (*Vues des Cordillères*), p. 144:
   See General 7.

3. Date 1870-74
   Stübel, A. 1897 (*Vulkanberge*), p. 145:
   'Gipfel des Quilindaña   4919 m
   Untere schneegrenze an der Nordseite im Toruno-huaico (mit Schutt
      bedeckter Schnee)   4364 m
   Fuss des Gletschers im Toruno-huaico   4470 m'

4. Date 1870-74
   Reiss, W. & Stübel, A. 1892-98 (*Hochgebirge*), pp. 160-162:
   'Nur an der Südseite giebt es ausgedehnte Schneefelder, sonst ragen überall die
   schwarzen Gesteinsfelsen aus dem weissen Schnee hervor. Die Schneegrenze mag etwa
   in 4600 Meter Höhe liegen, während die aus derselben hervortretenden Eismassen nur
   wenig weiter abwärts sich erstrecken. Die Gletscher hängen an den fast unersteiglichen
   Abstürzen der centralen Felspyramiden, ohne den Grund der Kesselthäler zu erreichen.
   Nur die von Zeit zu Zeit vom unteren Gletscherende abbrechenden Eis- und Schneemassen gelangen, über die Felswände stürzend, in den Thalgrund. Ich bestimmte das
   steil abgebrochene, untere Ende des Gletschers im Toruno-huaico zu 4470 m; Herr
   Dr. Stübel fand eine von Schutt bedeckte Schneemasse im Toruno-huaico in 4364 m;
   . . . Auch in den anderen, in die Seiten der Centralpyramide einschneidenden Kessel-
   thälern erreichen die Gletscher nirgends den Grund des Thales, sie bleiben stets hoch
   oben an den Felswänden hängen.'

5. Date 1877
   Dressel, L. 1877 (*Vulkane*), p. 450:
   See General 17.

6. Date 1903
   Meyer, H. 1907 (*Hochanden*), p. 265:
   '. . . öffnete sich für kurze Zeit der Ausblick auf die Nordnordwestseite des Quilindaña
   und zeigte dort in einem Kahr, das etwas höher liegt als das des Verde-cocha, ein kurzes
   Gletscherchen mit Eisbrüchen und davor unverhältnismässig hohe Endmoränenwälle
   . . . Auch auf der . . . felsigen Nordwestspitze des Quilindaña kam aus den Nebeln ein
   Kahr zum Vorschein. Es enthält jetzt zwar nur ein flaches Firnfleck, aber eine lange,
   von ihm sich vorstreckende alte Ufermoräne verrät, dass auch dieses Kahr einst glet-
   schergefüllt gewesen ist.'

7. Date 1903
   Meyer, H. 1907 (*Hochanden*), p. 269:
   'Auf der Nordostseite der Quilindañapyramide ist in etwa 4600 m Höhe ein breites
   Kahr mit einem kleinen Gletscher eingebetter, . . . Und auch sonst liegen auf seinem
   Gipfel und seinen Flanken mehrere Firnfelder mit Schründen und Eisbrüchen.'

8. Date 1903
   Meyer, H. 1907 (*Hochanden*), p. 272:
   (North side of Quilindaña)
   '. . . wir kamen . . . auf einen Hügel vor der Nordostseite des Filo de Verde-cocha, von
   wo aus ein anderes, kleineres, dem unsrigen ganz ähnliches Tal nach Nordnordwesten
   abzweigt. Wir sahen es oben von einem Kahr ausgehen, das an der Vereinigung des
   Filo de Verde-cocha mit der zentralen Quilindañapyramide ca. 4200 m hoch gelegen
   ist. Gegenwärtig birgt dieses Kahr wohl Schnee, aber keinen Gletscher.'

COTOPAXI

1. Date 1738-1740
   Reiss, W. & Stübel, A. 1892-98 (*Hochgebirge*), Vol. 2, p. 130:
   'Wir müssen also die Höhe des Cotopaxi vor den grossen Ausbrüchen des 18. Jahr-
   hunderts, etwa für die Jahre 1738-1740, nach Bouguer and La Condamine, zu 5750
   Meter annehmen.'

2. Date 1872
   Reiss, W. & Stübel, A. 1892-98 (*Hochgebirge*), Vol. 2, p. 130:
   'Wir können also die Höhe des Cotopaxi im Jahre 1872 zu 5944 Meter annehmen, dass
   heisst um 194 Meter höher, als Bouguer und La Condamine in den Jahren 1738-40
   den Berg gefunden hatten. In 130 Jahren hat in Folge der Ausbrüche, welche in dieser
   Zeit stattgefunden haben, der Cotopaxi an Höhe um 194 Meter zugenommen.'

3. Date 27 Nov. 1872
   Dietzel, K.N. (ed.) 1921 (Reiss, *Reisebriefe*), p. 176:
   'Zeltlager an der Schneegrenze, 4627 m.'

4. Date 1903
   Meyer, H. 1907 (*Hochanden*), pp. 250-251:

'Hier im Süden liegt die 'wirkliche' Schneegrenze bei 4730 m, im Westen bei 4850 m, im Norden bei 4900 m, aber im feuchten Osten bei 4550 m. W Reiss gibt für 1872 an: Süd- und Westseite 4630 m, Nordseite 4760 m, Ostseite 4550 m. Somit hat sich der Verlauf der Schneegrenze gegen den von Reiss vor 3 Jahrzehnten gemessnen um 100 bis 180 m auf der Süd-, West- und Nordseite aufwärts verschoben; nur im Osten ist sie gleich geblieben.'

RUMIÑAHUI

1. Date 1727-1767
   Velasco, J. de 1841-44 (*Historia*), Vol. 1, pp. 7-8:
See General 6.

2. Date 1802
   Humboldt, A. de 1810 (*Vues des Cordillères*), p. 44:
See General 7.

3. Date 1858
   Villavicencio, M. 1858 (*Geografía*), p. 66:
See General 8.

4. Date 23-25 Aug. 1870
   Dietzel, K.N. (ed.) 1921 (Reiss, *Reisebriefe*), p. 106:
'Der Rumiñahui . . . Seine wohl meist unersteiglichen Felsgipfel sind zwar nicht mit ewigem Schnee bedeckt, reichen aber doch nahe bis zur Schneegrenze, denn fast täglich fällt frischer Schnee bis weit herab am Gehänge.'

5. Date 1870-74
   Reiss, W. & Stübel, A. 1892-1898 (*Hochgebirge*), Vol. 1, p. 64:
'. . . aufragend bis zur Region des ewigen Schnees, bildet der Rumiñahui . . . einen der auffallendsten Berge in der Umgebung von Quito.'

6. Date 1877
   Dressel, L. 1877 (*Vulkane*), p. 450:
See General 17.

7. Date 1886
   Karsten, H. 1886 (*Géologie*), pp. 35-36:
See General 20.

8. Date 1892
   Wolf, T. 1892 (*Geografía y geología*), p. 76:
'. . . el volcán de Rumiñahui . . . tiene la altura de 4757 metros y no llega a la línea de la nieve perpétua.'

9. Date 1903
   Meyer, H. 1907 (*Hochanden*), p. 289:
See General 23.

SINCHOLAGUA

1. Date 15 March 1540
   González Rumazo, J. 1934a (*Cabildos de Quito*), Vol. 2, p. 105:
   See General 2.

2. Date 1541
   González Suarez, F. 1891 (*Ecuador*), Vol. 2, pp. 281-283:
   See General 3.

3. Date 12 March 1550
   González Rumazo, J. 1934b (*Cabildos de Quito*), Vol. 4, p. 311:
   See General 4.

4. Date 1727-1767
   Velasco, J. de 1841-44 (*Historia*), Vol. 1, pp. 7-8:
   See General 6.

5. Date 1870-74
   Stübel, A. 1897 (*Vulkanberge*), p. 146:
   'An ihrer Nordwestseite aber weist die Gipfelpyramide . . . eine breite, von zackigen Graten umgebene, stark vergletscherte Mulde auf (Schneegrenze = 4577 m).'
   Id., p.149:
   'Untere Schneegrenze auf der Nordseite des Sincholagua 4577 m.'

6. Date 1870-74
   Reiss, W. & Stübel, A. 1892-98 (*Hochgebirge*), Vol. 2, p. 65:
   'Auf einem ziemlich flachen . . . Unterbau, . . . erhebt sich schroff und steil die oberste mit Schnee und Eis bedeckte Felspyramide des Sincholagua.'

7. Date 1877
   Dressel, L. 1877 (*Vulkane*), p. 450:
   See General 17.

8. Date 1880
   Whymper, E. 1892 (*Travels*), pp. 160-161:
   'All the grass land was below, and we were confronted with crags, precipitous enough for any one, crowned by fields of snow and ice, the birthplace of a fine hanging-glacier which crept down almost perpendicular cliffs, clasping the rocks with its fingers and arms.'

PASOCHOA

1. Date 1802
   Humboldt, A. von 1874a (*Kosmos*), p. 381:
   See General 10.

2. Date 1858
   Villavicencio, M. 1858 (*Geografía*), p. 66:
   See General 11.

3. Date 1877
   Dressel, L. 1877 (*Vulkane*), p. 450:
See General 17.

4. Date 1886
   Karsten, H. 1886 (*Géologie*), pp. 35-36:
See General 20.

5. Date 1903
   Meyer, H. 1907 (*Hochanden*), p. 289:
See General 23.

ANTISANA

1. Date 1903
   Meyer, H. 1907 (*Hochanden*), pp. 334-335:
'Ich nenne den uns zunächstliegenden langen Eisstrom einfach Westgletscher, weil er
ziemlich in der Mitte der Westseite liegt; den nördlicheren, kürzeren aber Guagraialina-
   gletscher.'

SARA URCU

1. Date 1727-1767
   Velasco, J. de 1841-44 (*Historia*), Vol. 1, pp. 7-8:
See General 6.

2. Date 17 Oct. 1871
   Dietzel, K. N. (ed.) (Reiss, *Reisebriefe*), p. 120:
'Immerhin machte ich am folgenden Tage, trotz des entsetzlichen Wetters, einen Aus-
flug nach der Grenze des ewigen Schnees (4364 m). Der Gipfel ist höchstens 4800 m
hoch, und doch reichen die Gletscher bis 4176 m herab, an der Westseite über furchtbar
schroffe Wände wie ein Wasserfall herabkommend.'

3. Date 1870-74
   Reiss, W. & Stübel, A. 1892-98 (*Hochgebirge*), Vol. 2, p. 88:
'Am Sara-urcu fand ich die Schneegrenze in 4364 m Höhe, das Gletscher-Ende in
4242 m . . .'

4. Date 1870-74
   Stübel, A. 1897 (*Vulkanberge*), p. 108:
'Dafür zeugt besonders die Grenze des ewigen Schnees, welche am Sara-urcu bereits in
4364 m liegt . . .'

5. Date 17 April 1880
   Whymper, E. 1892 (*Travels*), p. 247:
'Saw that it was surrounded by glaciers on the South. The summit seemed to be a sharp
snow peak. This appearance we know was delusive.'
   Id., p. 249:
'. . . limiting the view to a few hundred yards around the summit, which was surrounded

116

by glaciers on all sides. We could not at any time see the full length of the large glacier on the west of Sara-urcu, or even across it. It appeared to bend round towards the north. The glaciers on the south side of the mountain are small. There was another one descending towards the north-east, which, so far as could be seen, was more considerable.'

6. Date 1892
   Wolf, T. 1892 (*Geografía y geología*), p. 406:
See General 22.

CAYAMBE

1. Date 1870-74
   Stübel, A. 1897 (*Vulkanberge*), p. 108:

'Westgipfel des Cayambe	5840 m
Ostgipfel	5556 m
Untere Schneegrenze auf der Nordostseite	4400 m
Untere Schneegrenze auf der Nordwestseite, bei Rumipungo	4672 m
Höchste schneefreie Stellen des Arenal von Rumipungu	4750 m
Fuss des Muyurcu-Gletschers an der Ostseite des Cayambe	4298 m
Fuss des Tarrugacorral-Gletschers	4134 m
Fuss des Yangurcal-Gletschers an der Nordseite des Cayambe	4510 m

2. Date 1870-74
   Reiss, W. & Stübel, A. 1892-98 (*Hochgebirge*), Vol. 2, p. 8:
'Bis beinah 4100 m ziehen aber einzelne Gletscher herab, und von 4600 bis 4700 m an ist der ganze Berg mit Schnee und Eis bedeckt . . .'

3. Date 1870-74
   Reiss, W. & Stübel, A. 1892-98 (*Hochbegirge*), Vol. 2, p. 54:
'Die Hauptfundpunkte dieser Gesteine sind die Moraine des Muyurcu-Gletschers und des Tarugacorral-Gletschers in einer Höhe von 4100-4500 m und an der Loma Rumipungu.'

4. Date April 1880
   Whymper, E. 1892 (*Travels*), pp. 232-233:
'Glaciers depart in all directions from the summit of Cayambe.'

APPENDIX III: DATA SUPPLIED TO WORLD GLACIER INVENTORY

Table A. Code numbers of Ecuadorian glaciers for World Glacier Inventory

The first three digits denote the independent political state (EC.); the fourth the South American continent (1); the fifth digit the first order drainage basin, that is discharge to the Pacific (N) and Atlantic (D) Oceans. The sixth through eighth digit are reserved to denote second, third and fourth order drainage basins, but are not used here. The tenth digit is used to denote the individual mountains, numbered from South to North. Digits eleven and twelve denote glaciers on individual mountains numbered as a rule clockwise from North, as shown in Maps 1-16.

Code for the glaciers of Ecuador:	EC.1x . . . / xxx		
Western Cordillera: (draining to Pacific)	EC.1N . . . / xxx		
COTACHACHI (4)	EC.1N . . . / 4xx		
glacier 1	EC.1N . . . / 4.1	glacier 2	EC.1N . . . / 4.2
ILINIZAS (3)	EC.1N . . . / 3xx		
glacier 1	EC.1N . . . / 3.1	glacier 6	EC.1N . . . / 3.6
2	EC.1N . . . / 3.2	7	EC.1N . . . / 3.7
3	EC.1N . . . / 3.3	8	EC.1N . . . / 3.8
4	EC.1N . . . / 3.4	9	EC.1N . . . / 3.9
5	EC.1N . . . / 3.5	10	EC.1N . . . / 310
CARIHUAIRAZO (2)	EC.1N . . . / 2xx		
glacier 1	EC.1N . . . / 2.1	glacier 6	EC.1N . . . / 2.6
2	EC.1N . . . / 2.2	7	EC.1N . . . / 2.7
3	EC.1N . . . / 2.3	8	EC.1N . . . / 2.8
4	EC.1N . . . / 2.4	9	EC.1N . . . / 2.9
5	EC.1N . . . / 2.5		
CHIMBORAZO (1)	EC.1N . . . / 1xx		
glacier 1	EC.1N . . . / 1.1	glacier 7	EC.1N . . . / 1.7
2	EC.1N . . . / 1.2	8	EC.1N . . . / 1.8
3	EC.1N . . . / 1.3	9	EC.1N . . . / 1.9
4	EC.1N . . . / 1.4	10	EC.1N . . . / 110
5	EC.1N . . . / 1.5	11	EC.1N . . . / 111
6	EC.1N . . . / 1.6	12	EC.1N . . . / 112

Table A (continued)

glacier 13	EC.1N . . . / 113	glacier 18	EC.1N . . . / 118
14	EC.1N . . . / 114	19	EC.1N . . . / 119
15	EC.1N . . . / 115	20	EC.1N . . . / 120
16	EC.1N . . . / 116	21	EC.1N . . . / 121
17	EC.1N . . . / 117	22	EC.1N . . . / 122

Eastern Cordillera: (draining to Atlantic)   EC.1D . . . / xxx

CAYAMBE (9)   EC.1D . . . / 9xx

glacier 1	EC.1D . . . / 9.1	glacier 11	EC.1D . . . / 911
2	EC.1D . . . / 9.2	12	EC.1D . . . / 912
3	EC.1D . . . / 9.3	13	EC.1D . . . / 913
4	EC.1D . . . / 9.4	14	EC.1D . . . / 914
5	EC.1D . . . / 9.5	15	EC.1D . . . / 915
6	EC.1D . . . / 9.6	16	EC.1D . . . / 916
7	EC.1D . . . / 9.7	17	EC.1D . . . / 917
8	EC.1D . . . / 9.8	18	EC.1D . . . / 918
9	EC.1D . . . / 9.9	19	EC.1D . . . / 919
10	EC.1D . . . / 910	20	EC.1D . . . / 920

SARA URCU (8)   EC.1D . . . / 8xx
No inventory was produced for this mountain.

ANTISANA (7)   EC.1D . . . / 7xx

glacier 1	EC.1D . . . / 7.1	glacier 10	EC.1D . . . / 710
2	EC.1D . . . / 7.2	11	EC.1D . . . / 711
3	EC.1D . . . / 7.3	12	EC.1D . . . / 712
4	EC.1D . . . / 7.4	13	EC.1D . . . / 713
5	EC.1D . . . / 7.5	14	EC.1D . . . / 714
6	EC.1D . . . / 7.6	15	EC.1D . . . / 715
7	EC.1D . . . / 7.7	16	EC.1D . . . / 716
8	EC.1D . . . / 7.8	17	EC.1D . . . / 717
9	EC.1D . . . / 7.9		

SINCHOLAGUA (6)   EC.1D . . . / 6xx

| glacier 1 | EC.1D . . . / 6.1 | glacier 3 | EC.1D . . . / 6.3 |
| 2 | EC.1D . . . / 6.2 | | |

COTOPAXI (5)   EC.1D . . . / 5xx

glacier 1	EC.1D . . . / 5.1	glacier 13	EC.1D . . . / 513
2	EC.1D . . . / 5.2	14	EC.1D . . . / 514
3	EC.1D . . . / 5.3	15	EC.1D . . . / 515
4	EC.1D . . . / 5.4	16	EC.1D . . . / 516
5	EC.1D . . . / 5.5	17	EC.1D . . . / 517
6	EC.1D . . . / 5.6	18	EC.1D . . . / 518
7	EC.1D . . . / 5.7	19	EC.1D . . . / 519
8	EC.1D . . . / 5.8	20	EC.1D . . . / 520
9	EC.1D . . . / 5.9	21	EC.1D . . . / 521
10	EC.1D . . . / 510	22	EC.1D . . . / 522
11	EC.1D . . . / 511	23	EC.1D . . . / 523
12	EC.1D . . . / 512		

QUILINDAÑA (4)	EC.1D . . . / 4xx		
glacier 1	EC.1D . . . / 4.1	glacier 2	EC.1D . . . / 4.2

CERRO HERMOSO (3) EC.1D . . . / 3xx
No inventory was produced for this mountain.

TUNGURAHUA (2)   EC.1D . . . / 2xx
No inventory was produced for this mountain.

EL ALTAR (1)	EC.1D . . . / 1xx		
glacier 1	EC.1D . . . / 1.1	glacier 6	EC.1D . . . / 1.6
2	EC.1D . . . / 1.2	7	EC.1D . . . / 1.7
3	EC.1D . . . / 1.3	8	EC.1D . . . / 1.8
4	EC.1D . . . / 1.4	9	EC.1D . . . / 1.9
5	EC.1D . . . / 1.5		

CUBILLIN (0)   EC.1D . . . / 0xx
No inventory was produced for this mountain.

SANGAY (0)   EC.1D . . . / 0xx
No inventory was produced for this mountain.

SOROCHE and
COLAY (0)   EC.1D . . . / 0xx
No inventory was produced for this mountain.

Table B. Characteristic glacier parameters. For classification code, see Table C.

Glacier number	Lat. S ° ′	Long. W ° ′	Total area $(10^4 m^2)$	Mean width $(10^2 m)$	Mean length $(10^2 m)$	Orien- tation	Highest elevation (m)	Lowest elevation (m)	Classifi- cation
CHIMBORAZO									
1	1 26.3	78 48.7	160	8	20	N	5,600+	4,700	400310
2	1 26.4	78 48.1	140	7	20	N	5,600+	4,700	400310
3	1 26.5	78 47.7	170	7	25	NE	5,600+	4,600	400310
4	1 27.0	78 47.4	160	7	23	NE	5,600+	4,600	400310
5	1 27.5	78 47.4	80	2	4	NE	5,600+	4,800	400310
6	1 27.7	78 47.3	120	5	20	NE	5,600	4,800	400310
7	1 28.2	78 47.2	120	5	20	E	5,600	4,800	400310
8	1 28.4	78 47.2	80	2	4	SE	5,600	4,800	400310
9	1 28.6	78 47.3	120	5	20	SE	5,600	4,800	400310
10	1 28.6	78 47.5	30	5	10	S	5,600	4,000	400310
11	1 28.6	78 47.6	30	5	10	S	5,400	4,800	400310
12	1 28.6	78 47.7	30	5	10	S	5,600	4,800	400310
13	1 28.7	78 47.9	80	4	20	S	5,600	4,900	400310
14	1 28.9	78 48.2	80	4	20	S	5,600	4,950	400310
15	1 28.9	78 48.6	80	4	20	S	5,600	4,900	400310
16	1 29.0	78 49.2	80	4	20	S	5,600	4,800	400310
17	1 28.9	78 49.2	100	5	20	SW	5,600	4,800	400310
18	1 28.6	78 50.0	70	3	20	SW	5,600	4,800	400310
19	1 28.1	78 50.5	130	6	23	SW	5,600	4,800	400310
20	1 27.9	78 50,7	140	7	24	W	5,600	4,750	400310
21	1 27.5	78 50.8	10	1	10	W	5,000	4,800	000020
22	1 27.0	78 49.8	14	7	24	NW	5,600	4,900	400310

Table B (continued)

Glacier number	Lat. S ° ′	Long. W ° ′	Total area $(10^4 m^2)$	Mean width $(10^2 m)$	Mean length $(10^2 m)$	Orientation	Highest elevation (m)	Lowest elevation (m)	Classification
CARIHUAIRAZO									
1	1 23.7	78 44.8	6	1	4	NE	4,700	4,550	700000
2	1 24.0	78 44.8	6	1	4	NE	4,700	4,550	700000
3	1 24.2	78 44.5	6	1	4	NE	4,700	4,550	700000
4	1 24.6	78 44.9	10	2	5	S	4,800	4,650	500310
5	1 24.5	78 45.0	10	2	5	S	4,800	4,650	500310
6	1 24.5	78 45.3	10	2	5	S	4,800	4,650	500310
7	1 24.2	78 45.7	10	2	5	W	4,900	4,650	403110
8	1 24.0	78 45.7	10	2	5	W	4,900	4,650	403110
9	1 23.9	78 45.7	10	2	5	W	4,900	4,650	403110
ILINIZAS									
1	0 39.3	78 42.6	10	2	5	NE	5,100	4,800	400310
2	0 39.6	78 42.6	6	2	4	E	5,100	4,800	400310
3	0 39.7	78 42.6	6	2	4	E	5,100	4,800	400310
4	0 39.8	78 42.6	6	2	4	E	5,100	4,800	400310
5	0 39.8	78 42.6	6	2	4	SE	5,100	4,800	400310
6	0 39.8	78 42.8	10	2	7	SW	5,100	4,800	400310
7	0 39.7	78 43.0	10	2	7	SW	5,100	4,800	400310
8	0 39.5	78 43.0	10	2	7	W	5,100	4,800	400310
9	0 39.2	78 42.9	10	2	5	N	5,100	4,800	400310
10	0 39.3	78 42.7	10	2	5	N	5,100	4,800	400310
COTACACHI									
1	0 22.3	78 20.6	3	1	3	W	4,850	4,750	200010
2	0.22.2	78 20.6	3	1	3	SW	4,850	4,750	200010
EL ALTAR									
1	1 39.2	78 25.2	20	3	6	S		4,400	600120
2	1 40.0	78 24.8	300	4	20	W		4,150	560320
3	1 40.6	78 25.4	20	5	5	N			660420
4	1 38.5	78 25.7	100	8	11	N			403100
5	1 38.6	78 24.7	120	9	12	N			403100
6	1 39.1	78 23.7	230	10	15	NE			403100
7	1 40.3	78 23.5	230	10	15	E			403100
8	1 41.1	78 23.7	230	10	15	SE			403100
9	1 41.4	78 25.0	230	10	15	S			403100
QUILINDAÑA									
1	0 46.7	78 19.3	3	1	3	S	4,700	4,600	500000
2	0 46.7	78 19.7	3	1	3	S	4,700	4,600	500000
COTOPAXI									
1	0 39.4	78 25.7		3	20	N		4,500	400310
2	0 39.5	78 25.5		3	20	N		4,500	400310
3	0 39.6	78 25.3		3	20	NE		4,500	400310
4	0 39.7	78 25.1		3	20	NE		4,500	400310
5	0 40.0	78 24.8		3	20	NE		4,500	400310
6	0 40.6	78 24.7		3	20	E		4,500	400310
7	0 41.2	78 24.7		3	20	E		4,500	400310
8	0.41.7	78 25.0		3	20	SE		4,500	400310
9	0 41.8	78 25.0		3	20	SE		4,500	400310
10	0 42.1	78 25.1		3	20	SE		4,500	400310
11	0 42.4	78 25.8		3	20	S		4,500	400310
12	0 42.3	78 26.1		3	20	S		4,500	400310
13	0 42.0	78 26.7		3	20	SW		4,700	400310
14	0 41.7	78 26.9		3	20	SW		4,700	400310
15	0 41.6	78 27.1		3	20	SW		4,700	400310

Table B (continued)

Glacier number	Lat. S ° ′	Long. W ° ′	Total area (10⁴m²)	Mean width (10²m)	Mean length (10²m)	Orientation	Highest elevation (m)	Lowest elevation (m)	Classification
16	0 40.4	78 27.2		3	15	W		4,700	400310
17	0 40.1	78 27.2		3	15	NW		4,700	400310
18	0 40.0	78 27.1		3	15	NW		4,700	400310
19	0 39.8	78 27.1		3	15	NW		4,700	400310
20	0 39.6	78 27.2		3	17	NW		4,550	400310
21	0 39.4	78 26.9		4	20	NW		4,500	400310
22	0 39.4	78 26.6		4	20	N		4,500	400310
23	0 39.1	78 26.2		4	22	N		4,500	400310
SINCHOLAGUA									
1	0 31.9	78 22.2	6	2	3	N	4,800	4,750	200000
2	0 32.2	78 22.1	6	2	3	S	4,800	4,700	200000
3	0 32.1	78 22.5	6	2	3	SW	4,800	4,700	200000
ANTISANA									
1	0 28.1	78 07.6	200	10	20	NE	5,400	4,450	400310
2	0 28.3	78 07.0	220	8	27	NE	5,400	4,200	400310
3	0 28.5	78 06.8	100	5	20	NE	5,400	4,400	400310
4	0 29.2	78 06.5	160	7	24	E	5,400	4,400	400310
5	0 29.6	78 06.6	160	7	24	SE	5,400	4,400	400310
6	0 30.1	78 07.2	160	7	24	S	5,400	4,350	400310
7	0 30.2	78 07.5	160	7	24	S	5,400	4,500	400310
8	0 30.2	78 08.0	160	7	24	S	5,400	4,500	400310
9	0 30.1	78 08.5	160	7	24	S	5,400	4,700	400310
10	0 30.0	78 08.8	160	7	24	SW	5,400	4,700	400310
11	0 29.8	78 09.0	160	7	24	SW	5,400	4,800	400310
12	0 29.5	78 09.4	200	8	25	W	5,400	4,800	400310
13	0 29.0	78 09.3	200	8	25	W	5,400	4,850	400310
14	0 28.6	78 09.2	180	7	24	W	5,400	4,850	400310
15	0 28.2	78 09.0	200	8	25	W	5,400	4,850	400310
16	0 27.8	78 08.5	170	7	22	N	5,400	4,550	400310
17	0 27.8	78 08.1	170	7	22	N	5,400	4,550	400310
CAYAMBE									
1	0 02.9N	77 59.8	150	6	25	N			400310
2	0 02.8	77 59.3	120	6	20	NE			400310
3	0 02.4	77 59.3	120	6	20	NE			400310
4	0 02.1	77 59.3	120	6	20	NE			400310
5	0 01.7	77 58.4	120	6	20	N			400310
6	0 01.7	77 58.1	120	6	20	NE			400310
7	0 01.2	77 58.0	120	6	20	NE			400310
8	0 01.0	77 57.7	120	6	20	NE			400310
9	0 00.4	77 56.7	120	6	20	NE			400310
10	0 00.1	77 56.8	120	6	20	E			400310
11	0 00.2S	77 56.9	120	6	20	SE			400310
12	0 00.5	77 58.1	120	6	20	S			400310
13	0 00.6	77 58.7	120	6	20	S			400310
14	0 00.6	77 59.0	120	6	20	S			400310
15	0 00.6	77 59.4	120	6	20	S			400310
16	0 00.4	78 00.2	120	6	20	S			400310
17	0 00.1	78 00.9	120	6	20	SW			400310
18	0 00.8	78 00.8	120	6	20	SW			400310
19	0 01.2	78 00.8	120	6	20	W			400310
20	0 03.1	78 00.2	120	6	20	N			400310

Table C. Classification according to World Glacier Inventory

	digit 1 primary classification	digit 2 form	digit 3 frontal characteristic	digit 4 longitudinal profile	digit 5 major source of nourishment	digit 6 activity of tongue
0	uncertain or misc.	uncertain or misc.	normal or misc.	uncertain or misc.	uncertain or misc.	uncertain
1	continental ice sheet	compound basins	Piedmont	even; regular	snow and/or drift snow	marked retreat
2	ice-field	compound basin	expanded foot	hanging	avalanche ice and/or avalanche snow	slight retreat
3	ice cap	simple basin	lobed	cascading	superimposed ice	stationary
4	outlet glacier	cirque	calving	ice-fall		slight advance
5	valley glacier	niche	confluent	interrupted		marked advance
6	mountain glacier	crater				possible surge
7	glacieret and snowfield	ice apron				known surge
8	ice shelf	group				oscillating
9	rock glacier	remnant				

APPENDIX IV: THE PLEISTOCENE CHANGES OF VEGETATION AND CLIMATE IN THE NORTHERN ANDES

T. VAN DER HAMMEN

Hugo de Vries Laboratorium, Afdeling Palynologie & Palaeoecologie, University of Amsterdam, Netherlands

Abstract. Palynological studies in the Northern Andes have shown a gradual upheaval of the Cordillera during the Late Pliocene and the creation of the high montane environment. A long sequence of glacial and interglacial periods has been recorded from the Pleistocene. The successive appearance of new taxa, by evolutionary adaptation from the local neotropical flora and from elements immigrated from the holarctic and austral-antarctic floral regions, can be followed step by step. For the Last Glacial to Holocene sequence the contemporaneity of the changes of temperature with those recorded from the northern temperate latitudes could be proved by ^{14}C dating. During the coldest part of the Last Glacial the tree line descended to c. 2000 m altitude, i.e. 1200-1500 m lower than where it lies today. During the period from c. 21,000 to c. 13,000 B.P. the climate was, moreover, much drier. Even taking the greater aridity into account, the lowering of the temperature during the coldest part of the Last Glacial may have been 6-7°C or more. The lowering of the temperature in the tropical lowlands during glacial times may have been c. 3°C. The temperature gradient must, therefore, have been steeper than it is today.

Introduction. It has been known for a considerable length of time that the Pleistocene glacials and interglacials changed repeatedly and greatly the face of the earth in the northern and temperate latitudes. This happened during at least the last two million years. It caused extinction, speciation and profound changes in the geographical distribution of plants and animals.

It is only rather recently that we learned that the so-called stable tropics, likewise, became subjected to drastic changes in climates which were apparently contemporaneous with those in the northern hemisphere.

In the last 20 years a considerable amount of data relevant to these changes in northern South America has become available, especially from the Northern Andes, but also from the tropical lowlands (van der Hammen, 1974). I will try to present here the broad outlines of these results, which are mainly based on pollen analyses, and partly still unpublished.

126

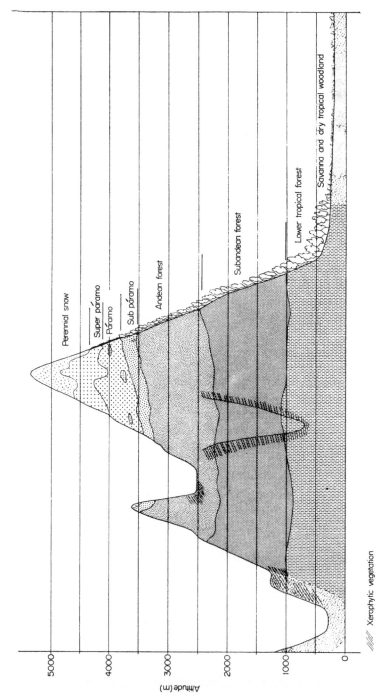

Figure 1. Present vegetation belts in the Cordillera Oriental (Colombia) shown schematically. The symbols are the same as those used in the diagrams.

The Northern Andes. The present vegetation belts (Figure 1). The Eastern Cordillera of the Northern Andes rise from tropical lowlands, where rain forests, savannas, or xerophytic vegetation types dominate. To the NE of this Cordillera lies the savanna area of the Llanos Orientales and the Orinoco, to the SE the rain forest. West of the Cordillera lies the Magdalena valley, the northern part of which bears rain forest and the southern part tropical xerophytic vegetation.

In the Eastern Cordillera the tropical belt extends from these lowlands to approximately 1000 m. At about this altitude several tropical taxa, such as the Bombacaceae, disappear, whereas several other ones are restricted to this belt or to a part of it (e.g. *Byrsonima, Iriartea, Mauritia* and *Spathiphyllum*).

The next altitudinal zone is that of the Subandean Forest, between c. 1000 and c. 2300 (−2500) m. Such genera as *Acalypha, Alchornea* and *Cecropia,* good pollen producers, are of frequent occurrence in this zone and do not extend beyond its upper limit. The same holds for many Palmae, *Hieronima, Ficus* and Malpighiaceae.

From 2300 (−2500) to 3200-3500 m the Andean Forest belt is present. In this belt forests of *Weinmannia* sp. div. and *Quercus* dominate. *Alnus, Myrica, Styloceras, Podocarpus, Clusia, Rapanea, Juglans, Ilex,* and *Hedyosmum* etc. are frequently present, although most of these genera are not restricted in their distribution to this belt.

The next higher belt is that of the high Andean dwarf forest and shrub formations, and the Subpáramo. It may be developed as a rather irregular belt, especially at its upper limit. Patches of this type of forest or shrub may be found at altitudes of up to 4000 m and over. The forest trees of the genera *Weinmannia* and *Quercus* are absent and the commonest woody taxa are various Compositae and Ericaceae, *Polylepis, Aragoa, Hypericum* etc. Some species of *Espeletia* may be present.

The proper Páramo belt extends from c. 3500 m up to 4000-4200 m. Open Andean grasslands may, however, be found from 3200 to 3300 m in the Subpáramo zone, patches of forest and shrub occurring in the higher grasslands at altitudes of up to 4000 m and over. Bogs and mires may be frequent. Apart from grasses and some sedges, the most characteristic elements are species of *Espeletia.* Amongst the herbs *Gentiana, Halenia, Valeriana, Bartschia, Geranium, Plantago, Ranunculus* and *Paepalanthus* may be mentioned.

The Super Páramo belt extends from 4000 to 4200 m upwards. Frost action on the soil is common here and the vegetation cover is incomplete to very scanty. *Espeletia* is mostly lacking. Characteristic elements are e.g. *Draba* sp. div. and *Senecio niveoaureus.* The proper nival zone, practically devoid of vegetation, extends from 4500 to 4800 m, or locally from somewhat higher altitudes upwards. The highest areas extending to c. 5500 m may be covered by snow and ice.

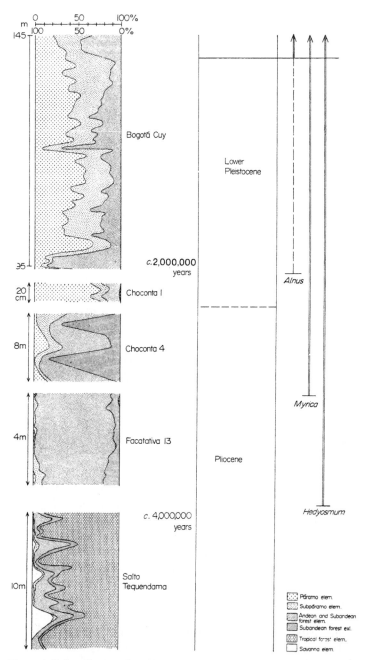

Figure 2. Pollen diagrams from the Pliocene and Lower Pleistocene of the high plain of Bogotá (Colombia), demonstrating the uplift of the area and the Early Pleistocene glacials and interglacials.

128

The ecological grouping of pollen types for the pollen diagrams. In order to make the pollen diagrams presented here easier to understand for non-palynologists, we have redrawn them as cumulative diagrams showing the percentage variation in time of ecological groups of pollen grains. These groups are the following:
1. Pollen from taxa common in the Páramo (dominating elements Gramineae).
2. Pollen from taxa common in the highest zone of the Andean Forest and shrub, or the Subpáramo, respectively.
3. Pollen from elements common in both the Andean and Subandean Forest.
4. Pollen from elements of the Subandean Forest.
5. Pollen from elements common in the tropical forest.
Although these types of diagrams give, generally speaking, a very clear picture of the changing vegetation, it will be clear that for a correct interpretation one must take into account both the present-day relation between vegetation and pollen rain in the area and the altitudinal range and ecology of the individual taxa.

The Pliocene, the upheaval of the Andes, and the early montane vegetation types (Figures 2 and 3). At the beginning of the Pliocene the area of the Eastern Cordillera lay mainly in the tropical belt. Folding had already taken place, so that hill ridges and low mountains, probably not exceeding 1000 m, were present. Rather extensive lowlands extended in broad synclinal basins, where fluvial and lacustrine sedimentation took place. This is proved by the pollen diagram of the lower part of the Tilatá formation (Figure 2, lower part), which diagram shows a complete dominance of tropical elements. The middle part of the Tilatá formation of later Pliocene age provided pollen diagrams showing an uplift of the basins in which sedimentation continued during this later stage of the Pliocene (Figure 2, middle part). The specific content shows that the dominating Subandean-Andean pollen group represents elements of the Subandean Forest, indicating an altitude of sedimentation of c. 1500 m and c. 2300 m respectively. When this elevation was reached, open high Andean vegetation, somehow comparable with the recent Páramo, must have been developed already on nearby higher mountains. During the Pliocene, *Hedyosmum,* and later *Myrica,* appeared for the first time. Sediments of the uppermost part of the formation already show a dominance of the primitive Páramo (Figure 2), although its present elevation is in the lower part of the Andean Forest belt. This means that the climate was considerably colder than it is today, and this part of the diagram must represent already a very early glacial, possibly of an age of some three million years.

There are clear indications that at this time the Andean Forest belt was not yet fully developed; it may have been narrower because the process of

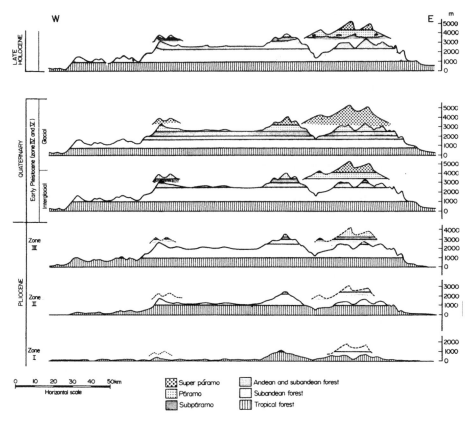

Figure 3. Sections through the Cordillera Oriental (Colombia) showing tentative
reconstruction of vegetation belts during the successive stages of uplift and during an
early Pleistocene interglacial and glacial. The uppermost section shows the present
situation. The main section is east-west at the latitude of Bogotá. The small sections
above each section are from higher areas farther to the north.

adaptation to the new biotopes had only just begun. Important taxa nowa-
days frequent in this belt, such as *Quercus* and *Alnus*, were absent at that
time. Similarly, the primitive Páramo vegetation is still very poor in species.
Apart from the dominant grasses, the earliest elements of this vegetation
include *Polylepis, Aragoa, Hypericum, Miconia,* Umbelliferae, *(Borreria,
Jussiaea, Polygonum), Valeriana, Plantago,* Ranunculaceae, *Myriophyllum*
and *Jamesonia.* Some of these elements were derived from the local flora,
whereas other ones must have been derived from founder species which
arrived at the newly created 'islands' of Páramo by long-distance dispersal.
The chances of such founders arriving and establishing themselves were certainly increased by the formation of the isthmus of Panama and by the

130

considerable enlargement of the Páramo islands (and the enlargement and displacement of their areas of origin in the Holarctic and Antarctic floral areas).

In Figure 3, the successive upheaval of the Cordillera and the creation of vegetation belts is shown diagrammatically (for further details, see Van der Hammen, Werner & van Dommelen 1973).

The Lower and Middle Pleistocene sequence (Figure 2, upper part and Figure 4). At the beginning of the Pleistocene, the principal upheaval of the area had ceased. Several hundreds of metres of lake sediments were deposited during the Pleistocene in the basin of the high plain of Bogotá (altitude c. 2580 m), providing us with a unique, very long and continuous palynological record of the changing vegetation and climate.

During the Lower Pleistocene several conspicuous, recurrent changes of the vegetation cover from Andean Forest to open Páramo (and *vice versa*) took place (Figure 2, upper part; Figure 3, upper part; and Figure 4, lower part). They were caused by the changes of climate from glacial to interglacial or the other way around, resulting in a downward and upward displacement of vegetation zones. At the same time at least the upper vegetation belts were gradually enriched by new taxa. In the Páramo belt *Geranium* and *Lycopodium* appear and somewhat later *Gunnera, Gentiana corymbosa* and *Lysipomia* sp. In the Andean Forest belt there appeared *Styloceras,* and somewhat later *Juglans* and Urticaceae. It seems as if *Alnus* appeared for the first time on the eastern slopes of the Eastern Cordillera during the later part of the Lower Pleistocene.

Our conclusion that the climate prevailing during the Early Pleistocene glacials was really much colder than it is today at the same elevation, is corroborated by the occurrence of simultaneous depositions of solifluction and fluvioglacial sediments.

At the beginning of the Middle Pleistocene (Figure 4, middle part) *Alnus* immigrated into the high plains. It had apparently become adapted to the ecological conditions of the Andean Forest belt and suddenly started to become a quantitatively important element (at least on wetter soil types).

The gradual enrichment of the flora continued (see Van der Hammen & Gonzalez, 1964), while the rhythm of glacial-interglacial displacement of vegetation belts continued likewise and resulted in the development of Páramo vegetation and Andean Forest on the mountain slopes surrounding the high plain, respectively.

Towards the end of the Middle Pleistocene *Quercus* made its first appearance in the Eastern Cordillera, as is manifest from the pollen diagrams (age: approximately 1,000,000 years ago). During the Upper Pleistocene it increases in quantitative importance, apparently becoming progressively adapted to the environmental conditions prevailing in the higher parts of

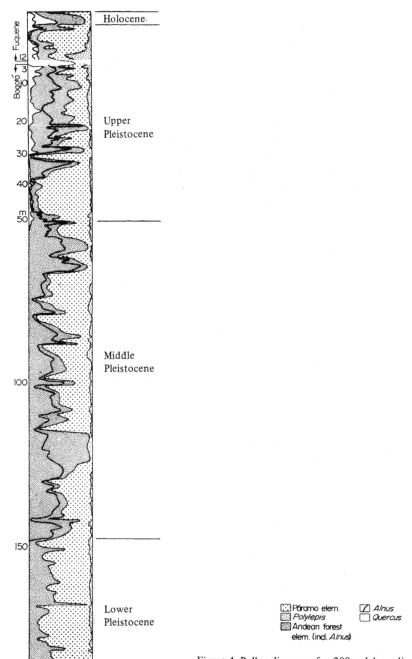

Figure 4. Pollen diagram of c. 200 m lake sediments from the high plain of Bogotá (section CUY); the uppermost part is from Fuguene.

Legend:

:::: Páramo elem.   ▨ Alnus
▦ Polylepis   ☐ Quercus
▩ Andean forest
elem. (incl. Alnus)

the Andean Forest belt (Figure 4, see also Van der Hammen et al., 1973).

The Upper Pleistocene: the last glacial-interglacial-glacial cycle. The Upper Pleistocene of the 'Sabana de Bogotá' (altitude c. 2580 m) was studied in greater detail (Van der Hammen & Gonzalez, 1960a). The terminal stage of its first glacial period apparently was very cold and dry; the pollen diagram (Figure 4) suggests the absolute dominance of open grass-páramo with such high-páramo species as *Malvastrum acaule*. Virtually no trace of forest elements remains.

During the last interglacial that probably started about 130,000 years ago, the area around the high plain is forested once more; the frequency of some of the elements from the uppermost Subandean Forest seems to indicate that the climate became even slightly warmer than it is at the present time.

The first part of the Last Glacial (the 'Early Glacial'), with a much wetter climate, shows several interstadials and stadials comparable to those of the northern temperate latitudes. During the forest phases of this Early Glacial, *Quercus* becomes a much more dominant element than before.

The following Pleniglacial was much colder and Páramo vegetation dominated. Elements of the uppermost forest and shrub and Subpáramo became more important during some of the minor 'interstadials' of the Middle Pleniglacial. The coldest phase of the Last Glacial started approximately 26,000 years ago. Around that time the Pleistocene lake of the 'Sabana de Bogotá' had dried up (see Figure 4). Sedimentation continued, however, in the lake of Fuquene on the next high plain (also at c. 2580 m), to the north of the plain of Bogotá, where we can follow the history until the present.

The pollen diagram from Laguna de Fuquene (Figure 5, Van Geel & Van der Hammen 1973) provides a fine and complete record of the vegetational history of the last 32,000 years. The results are summarized in Figure 6. From the pollen diagram we may deduce that fluctuations in the temperature and in humidity occurred. The first kind of fluctuation can be estimated by means of the displacement of vegetation belts, and the second can be deduced from the extension and retraction of the marshy zone of the hydrosere reflected in the pollen diagram. Approximately 30,000 years ago *Polylepis* scrub dominated completely in the area, its occurrence indicating conditions slightly above the limit of the proper Andean Forest. Shortly afterwards the climate became progressively colder so that open Páramo vegetation started to dominate.

About 21,000 years ago the climate became extremely cold and dry. The water in the lake dropped to a very low level and extreme Páramo conditions prevailed. This lasted till the beginning of the Late Glacial, 13,000 years ago.

The lowering of vegetation belts during this period was of the order of 1200-1500 m. Taking the influence of aridity into account, the average

133

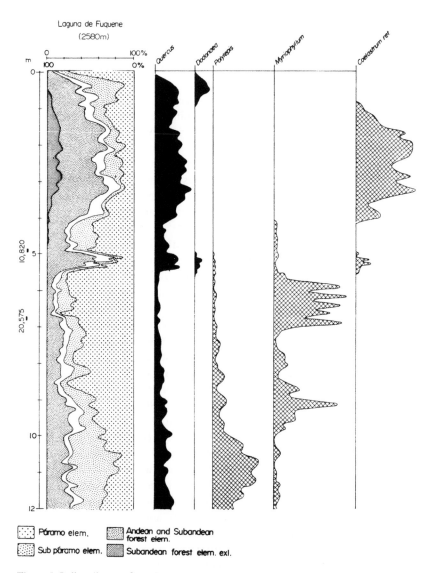

Figure 5. Pollen diagram from Laguna de Fuquene, Cordillera Oriental, Colombia (adapted from Van Geel & Van der Hammen 1973). The white area represents the percentage of *Alnus*.

annual temperature during this very cold Upper Pleniglacial may have been something like 8°C (and at least 6°C to 7°C) lower than it is today, and the annual precipitation may have been as low as 100-400 mm, which is less than half of the present value. We shall see presently that a lowering of the

134

tree line by about 1 500 m is confirmed by the pollen diagram from the Laguna de Pedro Palo at an elevation of 2000 m. During the Upper Pleniglacial the mountain glaciers extended downwards to an altitude of approximately 3000 m. This happened before c. 23,000 BP and after c. 28,000 BP. After that time, glaciers gradually retreat. Before that time there may have been a glacial extension reaching even further down, possibly of Middle Pleniglacial age. This has been found in recent studies in the Sierra Nevada del Cocuy and the highplain of Bogotá (van der Hammen c.s., 1980 and 1981). These relatively early maximal extensions of glaciers were apparently caused by an extremely wet Middle Pleniglacial and early Upper Pleniglacial climate, as compared with the very dry climate of the later Upper Pleniglacial. Prior to 21,000 BP there must have been a narrow zone of wet Páramo, between a comparatively high forest limit and low limit of glacier ice; glaciers and forest may have been locally in contact. After 21,000 BP there must have been a broad zone of dry Páramo, ice and forest being widely separated.

The Late Glacial and the Holocene. At present a considerable number of pollen diagrams covering the last 13,000 years have been drawn up from sediments in lakes situated at elevations between 2000 and 4000 m.

About 14,000-13,000 years ago the climate started to become less severe. This Late Glacial period lasted till approximately 10,000 years ago and exhibited several minor climatic fluctuations that could be correlated, by means of ^{14}C dating, with those of the northern temperate latitudes, and also with the climatic sequence in, e.g. tropical Africa.

In Fuquene (Figures 5 and 6) a major interstadial, the Guantiva interstadial, is represented. It lasted until c. 11,000 B.P. and was followed by the El Abra stadial that lasted until approximately 10,000-9,500 years ago, the beginning of the Holocene. During the Guantiva interstadial the area was forested. *Dodonaea,* a pioneer of bare soil, was abundant and the composition of the forest seems to indicate that the temperature was not much lower than it is at the present time. The incidence of the alga *Coelastrum reticulatum* corroborates this conclusion. During the El Abra stadial there was a considerable cooling of the climate, so that the site was near the forest limit. A striking fact is the sudden rising of the lake level at the beginning of the Guantiva interstadial, which indicates a much wetter climate than before. After a minor lowering of the water table during the El Abra stadial, the water in the lake again rose to a slightly higher level than today by the beginning of the Holocene.

In the higher mountains the glaciers had retreated with fluctuations since their maximum extension. During the Guantiva interstadial they had retreated to altitudes above 4000 m to descend again to c. 3900 m during the El Abra stadial.

135

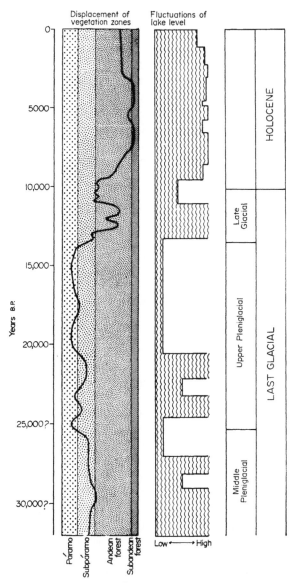

Figure 6. Displacement of vegetation zones and fluctuations of lake level, Laguna de Fuquene (see diagram Figure 5). (Adapted from Van Geel & Van der Hammen 1973).

During the Holocene the climate in the Fuquene area became even warmer than it is today during the 'hypsothermal'. Elements from the uppermost Subandean Forest *(Cecropia, Acalypha)* could even grow in the area at altitudes several hundreds of metres above their present upper limit of

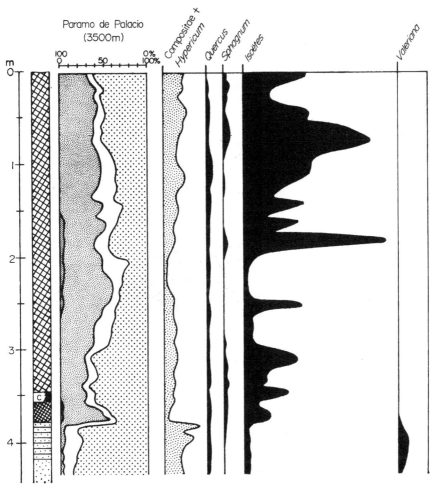

Figure 7. Pollen diagram of the Late Glacial and Holocene from Páramo de Palacio (Cordillera Oriental, Colombia). (Adapted from Van der Hammen & Gonzalez 1960b). The stratigraphical column indicates from top to bottom: detritus gyttja, volcanic ash, detritus gyttja, sandy clay and clayey sand. In this diagram certain elements of the Sub-páramo group (*Hypericum,* Compositae) were not included in the pollen total. The white area represents the percentage of *Alnus.*

occurrence. It seems, therefore, as if the annual temperature was about $2°C$ higher than today. The forest elements in question disappeared again about 3000 years ago when the temperature fell again to the present-day average.

An example of the succession at higher elevation is provided by the diagram from Páramo de Palacio (c. 3500 m), NE of Bogotá (Figure 7) (see Van der Hammen & Gonzalez 1960b). The site lies at 200-300 m above the present-day limit of the *Weinmannia* forest. The Late Glacial clayey sedi-

137

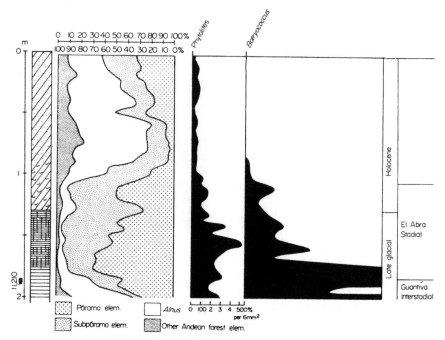

Figure 8. Pollen diagram of the Late Glacial and Holocene from El Abra (Sabana de Bogotá). The stratigraphical column indicates from top to bottom: dark partly sandy soil, black humic clay and grey lake clay. (Adapted from Schreve-Brinkman, 1978).

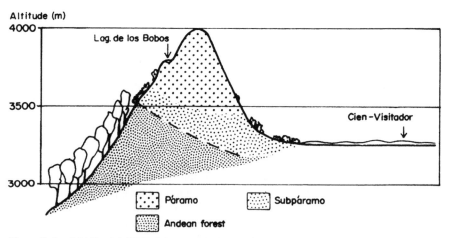

Figure 9. Section through part of the Cordillera Oriental in the area of Páramo de Guantiva (Cordillera Oriental, Colombia); the localities of the diagrams of Figures 10 and 11 are indicated.

138

ments, deposited after the retraction of the glaciers from the lake area, show fluctuations in the pollen diagram similar to those noted in the Late Glacial of Fuquene. During the Guantiva interstadial the forest limit seems to have been even slightly higher than where it lies today, although the composition of this forest seems to be somewhat different from the present, with Urticaceae being much more frequent. During the El Abra stadial the forest limit descended again, but during the Holocene hypsothermal this limit lay at an altitude several hundreds of metres higher than the present one, and the area around the lake must have been forested. At about 3000 years ago a lowering of the forest limit took place which persists until today.

Another Late Glacial-Holocene sequence is represented in a pollen dia-

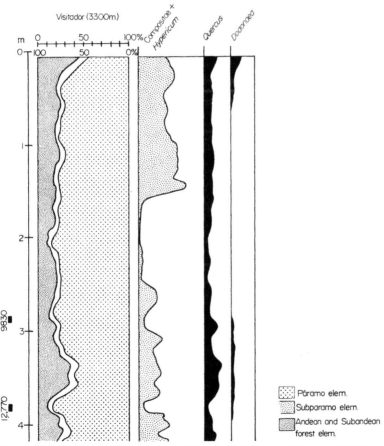

Figure 10. Pollen diagram from the Late Glacial and Holocene of Cienaga del Visitador (see Figure 9). The elements of the Subpáramo group Compositae and *Hypericum* are not included in the pollen total. The white area represents the percentage of *Alnus* in the pollen total. (Adapted from Van der Hammen & Gonzalez 1965.)

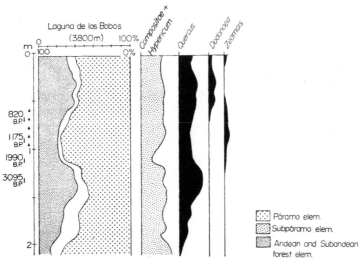

Figure 11. Pollen diagram from the Late Holocene of Laguna de Los Bobos (see Figure 9). The elements of the Subpáramo group Compositae and *Hypericum* are not included in the pollen total. The white area represents the percentage of *Alnus* in the pollen total. (Adapted from Van der Hammen 1962.)

gram from the El Abra valley in the high plain of Bogotá (Figure 8) (compare Schreve-Brinkman, 1978). The diagram starts with lake sediments (see the curve of the alga *Botryococcus*) of Guantiva interstadial age; the climate was wet and *Alnus* carr must have been abundant in the area. The latest part of this interstadial is dated here as 11,200 B.P. The climate became drier at the beginning of the El Abra stadial and the local lake became a marsh. The vegetation in the area became dominated by low forest and grassland, the presence of Cactaceae corroborating the palynological indications pointing to a relatively dry climate. During the Holocene forest vegetation dominated.

From the area of Páramo de Guantiva in the western part of the Cordillera Oriental, two pollen diagrams are presented to show the influence of local climatic conditions (Van der Hammen 1962; Van der Hammen & Gonzalez 1965). Figure 9 shows these conditions in an E-W transect. The western slopes of the mountains fall off steeply to much more low-lying areas and have a much wetter climate, the upper limit of the oak forest lying at c. 3500 m. Behind these mountains is shown a flat area at c. 3300 m that lies in rain shadow and supports only small patches of forest. The pollen diagram from this drier area (Figure 10) shows in its lower part a dated Late Glacial with the Guantiva interstadial. The climate was apparently much wetter during the interstadial and a much larger area was under forest than today. Trees disappeared from the area by the beginning of the cooler and drier El Abra

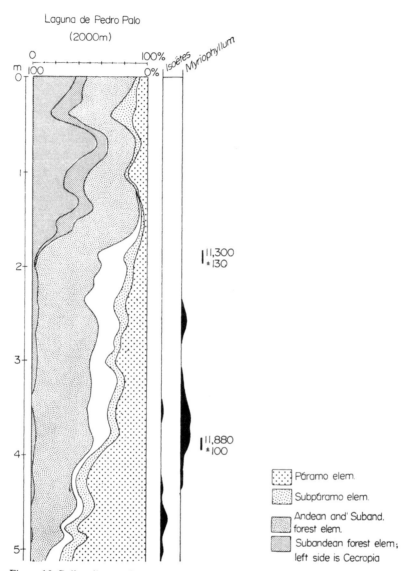

Figure 12. Pollen diagram from the Late Glacial and Holocene of Laguna de Pedro Palo (Cordillera Oriental, Colombia). (Adapted from Van der Hammen, in preparation). The white area represents the percentage of *Alnus* in the pollen total.

stadial. When at the beginning of the Holocene the climate became warmer again, the area did not become reforested, apparently because of the low amount of precipitation.

The pollen diagram from a small lake on the western slopes of these mountains show the conditions on the western wet slopes during the late Holocene

142

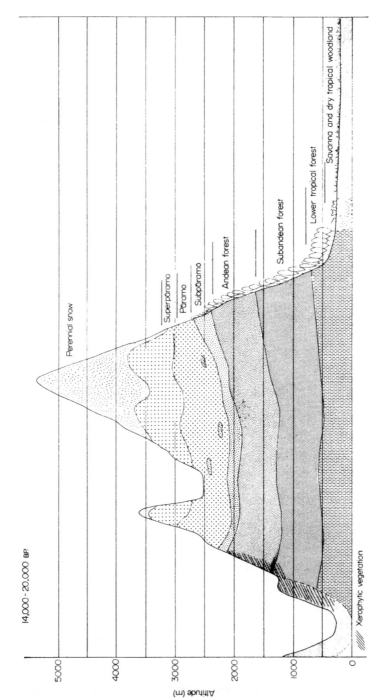

Figure 13. Tentative reconstruction of vegetational zones in the Eastern Cordillera during the Last Glacial maximum (compare with Figure 1).

(Figures 9 and 11). Even at an elevation of 3800 m the percentage of forest elements is much higher here than at the former site. A cooling is recorded at c. 3000 years ago.

From a lake (Laguna de Pedro Palo, W of the Sabana de Bogotá) at c. 2000 m, a most elucidating diagram of the Late Glacial was obtained (Figure 12; van der Hammen, in preparation). The site is in the upper part of the Andean Forest belt. Sedimentation started in the early Late Glacial or Late Pleniglacial. The area subsequently became covered with open grassland vegetation, and the forest limit must clearly have lain below 2000 m. The presence of grains of *Isoetes* and of *Myriophyllum* (and of other, non-aquatic, herbs), which taxa are today mainly found above 3000 m corroborates the conclusion that this open vegetation resembled Páramo vegetation very much, so that the timber line must have been at least 1300 (probably 1500) m lower than today. Some 12,000 years ago the area became invaded by Andean Forest, and towards the end of the Guantiva interstadial by Subandean Forest.

Conclusions (Figure 13). The tropical-montane, northern Andean climatic belts came into being during the Late Pliocene upheaval of the Cordillera. These newly created belts gradually became populated by processes of evolutionary adaptation of elements from the local Neotropical flora and by the arrival of elements which immigrated from the Holarctic and Antarctic floral areas. This process continued during the entire Pleistocene. In the Andean Forest belt local elements are frequent, but *Weinmannia* originally came from the south and *Myrica, Alnus* and *Quercus* from the north.

In the transitional and Subpáramo forest and scrub local genera are still abundant, but elements from more remote areas gradually became more frequent to reach their highest frequencies of occurrence in open Páramo vegetation (*Gentiana, Bartschia, Valeriana, Draba, Hypericum, Berberis* etc. from the Holarctic; *Muehlenbeckia, Acaena, Azorella* etc. from the Antarctic). The most characteristic Páramo genus, *Espeletia,* is of local Andean origin, however, and other endemics are, e.g. *Aciachne, Distichia, Puya* and *Rhizocephalum.*

The Pleistocene shows a sequence of several glacials and interglacials comparable with those of the northern hemisphere and the contemporaneity of the changes of temperature could be substantiated for the last 50,000 years, i.e. for the period within the reach of reliable ^{14}C dating.

The depression of the Andean Forest limit (presently between c. 3200 and 3500 m) was of the order of 1200-1500 m. Although this limit must have been as irregular as it is today owing to local climatic and microclimatic conditions, it seems as if a probable average value of 2000 m for its altitudinal position during periods of maximum glaciation can be considered to be established. Under extreme arid conditions the lowermost limit of dry Páramo vegetation in the Eastern Cordillera today is at 3000 m. If such conditions prevailed during the Upper Pleniglacial, we still have to accept a lowering of temperature of 6-7°C to explain the total depression of the altitudinal forest limit.

During the coldest period of the Last Glacial (later Upper Pleniglacial), the climate on the high plains was much drier than today, and for the areas of Fuquene the annual precipitation may have been less than half that at present. As we shall see presently, the temperature in the tropical lowlands during glacial time may only have been about 3°C lower than today; this means that the temperature gradient in the Northern Andes was much steeper than it is at present.

Major extension of glaciers took place before c. 23,000 B.P. (probably in the period between 44,000 and 24,000 B.P.), during very wet and relatively cold climatic conditions and a narrow zone of wet Páramo developed. Between 21,000 and 14,000 B.P., the glaciers retired strongly during very dry and cold conditions, and a broad zone of dry Páramo vegetation developed.

During certain intervals of glacial time the surface area occupied by Páramo vegetation was a multiple of its present extension; many now isolated 'islands' were in former times linked together. The Páramos of Cocuy and Sumapaz formed then part of a large and continuous area of Páramo that covered the entire central portion of the Cordillera Oriental. The Superpáramos of these two areas were never in direct contact with one another, however. This seems to be reflected in the relatively high degree of endemism, especially in the Superpáramo of the Sierra Nevada del Cocuy.

The distribution patterns in the Northern Andes may be partly explained by long-distance dispersal, partly by the erstwhile continuity of areas of Páramo during glacial times. Conditions for immigration by long-distance dispersal (or from one 'island' to another) were certainly much more favourable for elements of the high-Andean vegetation groups during glacial times. Speciation of both the local and the alien elements must have been stimulated by the successive periods of separation and union of populations.

References

Schreve-Brinkman, E.J., 1978. A palynological study of the Upper Quaternary sequence in the El Abra corridor and rock shelters. *Palaeogeog. Palaeoclim. Palaeoecol.* 25: 1-109.
Van der Hammen, T. 1962. Palinologia de la region de 'Laguna de los Bobos'. *Revta Acad. colomb. Cienc. exact. fis. nat.* 11, No. 44.
Van der Hammen, T. 1974. The Pleistocene changes of vegetation and climate in tropical South America. *Journ. Biogeography* 1:3-26.

Van der Hammen, T. (in preparation). Late Glacial and Holocene vegetation history of Laguna de Pedro Palo.

Van der Hammen, T. & E. Gonzalez 1960a. Upper Pleistocene and Holocene climate and vegetation of the 'Sabana de Bogotá' (Colombia, South America). *Leid. geol. Meded.* 25:126-315.

Van der Hammen, T. & E. Gonzalez 1960b. Holocene and Late Glacial climate and vegetation of Páramo de Palacio (Eastern Cordillera, Colombia, South America). *Geologie mijnb.* 39:737-746.

Van der Hammen, T. & E. Gonzalez 1964. A pollen diagram from the Quaternary of the Sabana de Bogotá (Colombia) and its significance for the geology of the Northern Andes. *Geologie mijnb.* 43:113-117.

Van der Hammen, T. & E. Gonzalez 1965. Late Glacial and Holocene pollen diagram from 'Cienaga del Visitador' (dept. Boyacá, Colombia). *Leid. geol. Meded.* 32:193-201.

Van der Hammen, T., J.H. Werner & H. van Dommelen 1973. Palynological record of the upheaval of the Northern Andes: a study of the Pliocene and Lower Quaternary of the Colombian Eastern Cordillera and the early evolution of its high-andean biota. *Palaeogeog. Palaeoclim. Palaeoecol.* 16:1-24.

Van der Hammen, T., H. Dueñas & J.C. Thouret 1980. Guia de excursión-Sabana de Bogota. Primer seminario sobre el Cuaternario de Colombia. *Bogotá*: 49 pp.

Van der Hammen, T., J. Barelds, H. de Jong & A.A. de Veer 1981. Glacial sequence and environmental history in the Sierra Nevada del Cocuy. *Palaeogeog. Palaeoclim. Palaeoecol.* 32 (1980/1981): 247-340.

Van Geel, B. & T. van der Hammen 1973. Upper Quaternary vegetational and climatic sequence of the Fuquene area (Eastern Cordillera, Colombia). *Palaeogeog. Palaeoclim. Palaeoecol.* 14:9-92.

REFERENCES

Allison, I. & P. Kruss 1977. Estimation of recent climatic change in Irian Jaya by numerical modelling of its tropical glaciers. *Arctic and Alpine Research 9:* 49-60.

Andrade Marín, L. 1936. *Viaje a las misteriosas montañas de Llanganati, 1933-1934.* Imprenta Mercantil, Quito. 1936, 239pp.

Atkinson, G. & J.C. Sadler 1970. *Mean cloudiness and gradient-level wind charts over the tropics.* USAF, Air Weather Service Technical Report No. 215.

Baker, B.H. 1967, *Geology of the Mount Kenya area.* Geological Survey of Kenya, Report No. 79, 78pp.

Bartels, G. 1970. *Geomorphologische Höhenstufen der Sierre Nevada de Santa Marta (Kolumbien).* Giessener Geographische Schriften, Vol. 21, 56pp.

Blomberg, R. 1952. *Ecuador, Andean mosaic.* Gebers, Stockholm, 320pp.

Bonifaz, E. 1971. Extractos de los libros del Cabildo de Quito, 1534-1657, pp. 124-180 in: *Museo Histórico, Organo del Archivo Municipal de Quito,* Vol. 7, No. 51.

Bonifaz, E. 1978. *Obsidianas del paleo-Indio de la región del Ilaló.* Quito, 105pp.

Bouguer, P. 1749. *La figure de la terre.* Jombert, Paris, 396pp.

Boussingault, J.P. 1835a. Ascension au Chimborazo executée le 16 décembre 1831. *Annales de Chimie et de Physique 58:* 150-180.

Boussingault, J.B. 1835b. Versuch einer Ersteigung des Chimborazo, unternommen am 16. December 1831. *Annalen der Physik und Chemie 34:* 193-219.

Boussingault, J.P. 1841. Rapport sur les travaux géographiques et statistiques exécutés dans la république de Venezuela. *Comptes Rendues de l'Académie des Sciences, Paris 12:* 462-479.

Boussingault, J.B. 1849. *Viajes científicos a los Andes ecuatoriales.* Libreria Castellana, Paris, 322pp.

Boussingault, J.B. 1851. *Économie rurale, considerée dans ses rapports avec la chimie, la physique et al métérologie.* 2nd ed., Vol. 2. Béchet Jeune, Paris, 767pp.

Bristow, C.R. 1973. *Guide to the geology of the Cuenca basin, Southern Ecuador.* Ecuadorian Geological and Geophysical Society, Quito, 54pp.

Broggi, J.A. 1943. La desglaciación actual de los Andes del Peru. *Bol. Soc. Geol. del Peru, 14 & 15:* 59-90.

Caukwell, R.A. & S. Hastenrath 1977. A new map of Lewis Glacier, Mount Kenya. *Erdkunde 31:* 85-87.

Caviedes, C.N. & R. Paskoff 1975. Quaternary glaciations in the Andes of North-Central Chile. *J. Glaciol. 14:* 155-170.

Chirriboga, G. 1969. *Libro de cabildos de la ciudad de Quito, 1650-57;* Publicaciones del Archivo Municipal, Vol. 33, 574pp.

Covey, D.L. & S. Hastenrath 1978. The Pacific El Niño phenomenon and the Atlantic circulation. *Mon. Wea. Rev. 106:* 1280-1287.

Clapperton, C.M. 1972. The pleistocene moraine stages of West-Central Peru. *J. Glaciol. 11:*255-263.

Dietzel, K.H., ed. 1921. *Wilhelm Reiss, Reisebriefe aus Südamerika 1868-1876.* Wissenschaftliche Veröffentlichungen der Gesellschaft für Erdkunde zu Leipzig, Vol. 9. Duncker & Humblot, München-Leipzig, 232pp.

Dirección General de Geología y Minas, Ecuador 1974. *Mapa geológico del Ecuador, escala 1:50,000, hoja 73 NW, Azogues.* Quito.

Dirección General de Geología y Minas, Ecuador 1975. *Mapa geológico del Ecuador, escala 1:100,000, hoja 72, Cañar.* Quito.

Downie, C. & P. Wilkinson 1972. *The geology of Kilimanjaro.* Department of Geology, University of Sheffield, 253pp.

Dressel, L. 1877. Die Vulkane Ecuadors und der jüngste Ausbruch des Cotopaxi. *Stimmen aus Maria-Laach 13:*446-464; 551-570.

Dressel., L. 1879. Erinnerungen aus Ecuador. *Stimmen aus Maria-Laach 16:*190-205; 462-479.

Dressel, L. 1880. Durch die Paramos zum äquatorialen Hochwald. *Stimmen aus Maria-Laach 18:*353-374; 498-515.

Eichler, A. 1952. *Nieve y selva en Ecuador.* Editorial Bruno Moritz, Quito-Guayaquil, 132pp.

Fernandez de Navarrete, M. 1829. *Colección de los viajes y descubrimientos que hicieron por mar los Espanoles,* Vol. 3. Imprenta Real, Madrid.

Furrer, G. & K. Graf 1978. Die subnivale Höhenstufe am Kilimandjaro und in den Anden Boliviens und Ecuadors. *Erdwissenschaftliche Forschung 11:*441-457.

Galloway, R.W., G.S. Hope, E. Loeffler & J.A. Peterson 1973. Late Quaternary glaciation and periglacial phenomena in Australia and New Guinea, pp. 125-128. In: E. van Zinderen-Bakker, ed., *Palaeoecology of Africa 8.* Rotterdam.

Gansser, A. 1955. Ein Beitrag zur Geologie und Petrographie der Sierra Nevada de Santa Marta (Kolumbien, Südamerika). *Schweizerische Mineralogische und Petrographische Mitteilungen 35:*209-278.

Garcés, J.A. 1934a. *Oficios o cartas al cabildo de Quito por el Rey de Espana o el Virrey de Indias, 1552-1568.* Publicaciones del Archivo Municipal, Vol. 5, 648pp.

Garcés, J.A. 1934b. *Libro de cabildo de Quito, 1573-74.* Publicaciones del Archivo Municipal, Vol. 6, 313pp.

Garcés, J.A. 1935. *Libro de cabildo de Quito, 1575-76.* Publicaciones del Archivo Municipal, Vol. 8, 341pp.

Garcés, J.A. 1937-40. *Libro de cabildos de la cuidad de Quito, 1597-1603.* Publicaciones del Archivo Municipal, Vols. 13 and 16, 417 and 413pp.

Garcés, J.A. 1941a. *Libro de cabildos de la ciudad de Quito, 1593-97.* Publicaciones del Archivo Municipal, Vol. 17, 450pp.

Garcés, J.A. 1941b. *Libro de proveimientos de tierras, cuadras, solares, aguas, etc. por los cabildos de la ciudad de Quito, 1583-1594.* Publicaciones del Archivo Municipal, Vol. 18, 248pp.

Garcés, J.A. 1944. *Libro de cabildos de San Francisco de Quito, 1603-10.* Publicaciones del Archivo Municipal, Vol. 20, 572pp.

Garcés, J.A. 1946-47. *Colección de documentos sobre el obispado de Quito, 1546-83.* Publicaciones del Archivo Municipal, Vols. 22 and 24, 601 and 590pp.

Garcés, J.A. 1953. *Descubrimiento del Rio de Orellana.* Imprenta Municipal, Quito, 149pp.

Garcés, J.A. 1955. *Libro de cabildos de la ciudad de Quito, 1610-16.* Publicaciones del Archivo Municipal, Vol. 26, 600pp.

Garcés, J.A. 1960. *Libro de cabildos de la ciudad de Quito, 1638-1646.* Publicaciones del Archivo Municipal, Vol. 30, 467pp.

Gerth, H. 1955. *Bau der südamerikanischen Kordillere,* Vol. 2. Gebrueder Borntraeger, Berlin, 264pp.

Giegengack, R. & R.I. Grauch 1973. Quaternary geology of the Central Andes, Venezuela, pp. 38-93, In: *Excursión No. 1, II Congreso Latinoamericano de Geología,* 38-93.

Gonzalez, E., Th. van der Hammen & R.F. Flint 1965. Late quaternary glacial and vegetational sequence in Valle de Lagunillas, Sierra Nevada del Cocuy, Colombia. *Leidse Geologische Mededelingen 32:* 157-182.

González Rumazo, J. 1934a. *Libro primero de cabildos de Quito (1534-43),* 2 vols. Publicaciones del Archivo Municipal, Vols. 1 and 2, 511 and 396pp.

González Rumazo, J. 1934b. *Libro segundo de cabildos de Quito (1544-51),* 2 vols. Publicaciones del Archivo Municipal, Vols. 3 and 4, 354 and 428pp.

Gonzalez Suarez, F. 1892-1903. *Historia del Ecuador,* 7 vols. Imprenta del Clero, Quito, 318, 463, 480, 487, 532, 261 and 153pp.

Graf, K. 1976. Zur Mechanik von Frostmusterungsprozessen in Bolivien und Ecuador. *Zeitschrift für Geomorphologie 20:* 417-447.

Grosser, P. 1905. Reisen in den ecuatorianischen Anden. *Sitzungsberichte der Niederrheinischen Gesellschaft für Natur- und Heilkunde zu Bonn 1:* 6-16.

Hastenrath, S. 1966. On general circulation and energy budget in the area of the Central American Seas. *J. Atm. Sci. 23:* 694-712.

Hastenrath, S. 1967. Rainfall distribution and regime in Central America. *Archiv Meteor. Geophys. Bioklim, Ser. B, 15:* 201-241.

Hastenrath, S. 1973. On the pleistocene glaciation of the Cordillera de Talamanca, Costa Rica. *Zeitschrift für Gletscherkunde und Glazialgeologie 9:* 105-121.

Hastenrath, S. 1974a. Spuren pleistozäner Vereisung in den Altos de Cuchumatanes, Guatemala. *Eiszeitalter und Gegenwart 25:* 25-34.

Hastenrath, S. 1974b. Glaziale und periglaziale Formbildung in Hoch-Semyen, Nord-Äthiopien. *Erdkunde 78:* 176-186.

Hastenrath, S. 1975. Glacier recession in East Africa, pp. 135-142. In: *Proceedings of WMO/IAMAP Symposium on long-term climatic variations, Norwich, England, 18-23 Aug. 1975.* WMO No. 421.

Hastenrath, S. 1976a. Variations in low-latitude circulation and extreme climatic events in the tropical Americas. *J. Atm. Sci. 33:* 202-215.

Hastenrath, S. 1976b. Marine climatology of the tropical Americas. *Archiv Meteor. Geophys. Bioklim., Ser. B, 24:* 1-24.

Hastenrath, S. 1977a. On the upper-air circulation over the equatorial Americas. *Archiv Meteor. Geophys. Bioklim., Ser. A, 25:* 309-321.

Hastenrath, S. 1977b. Pleistocene mountain glaciation in Ethiopia. *J. Glaciol. 16:* 309-313.

Hastenrath, S. 1978a. Heat budget measurements on the Quelccaya Ice Cap, Peruvian Andes. *J. Glaciol. 20:* 85-97.

Hastenrath, S. 1978b. On modes of tropical circulation and climate anomalies. *J. Atm. Sci. 35:* 2222-2231.

Hastenrath, S. & R.A. Caukwell 1979. Variations of Lewis Glacier, Mount Kenya, 1974-78. *Erdkunde 33,* 292-297.

Hastenrath, S. & P. Guetter 1978. A contribution to the surface circulation over South America. *Archiv Meteor. Geophys. Bioklim., Ser. B, 26:* 97-103.

Hastenrath, S. & L. Heller 1977. Dynamics of climatic hazards in Northeast Brazil. *Quart. J. Roy. Meteor. Soc. 103:* 75-90.

Hastenrath, S. & P. Lamb 1977. *Climatic atlas of the tropical Atlantic and Eastern Pacific Oceans.* University of Wisconsin Press.

Heine, K. 1975. *Studien zur jungquartären Glazialmorphologie Mexikanischer Vulkane.*

Mexiko Projekt DFG, Vol. 7. Steiner Verlag, Wiesbaden, 178pp.

Heinzelin, J. de 1962. Carte des extensions glaciaires du Ruwenzori (versant Congolais). *Biul. Periglacjalny 11:*133-139.

Herd, D.G. & C.W. Naeser 1974. Radiometric evidence for Pre-Wisconsin glaciation in the Northern Andes. *Geology 2:*603-604.

Herrera y Tordesillas, A. de 1934. *Historia general de los hechos de los Castellanos en las Islas y Tierrafirme del Mar Océano.* Vol. 1: *Descripción de las Indias Occidentales.* Academia de la Historia, Madrid.

Hope, G.S., J.A. Peterson, U. Radok & I. Allison 1976. *The Equatorial glaciers of New Guinea. Results of the 1971-73 Australian Universities expeditions to Irian Jaya.* A.A. Balkema, Rotterdam, 244p.

Humboldt, A. de 1810. *Vues des Cordillères et monumens des peuples indigènes de l'Amerique,* 2 vols. Schoell, Paris.Vol. 1, 350pp, Vol. 2, pictures.

Humboldt, A. von 1853. *Kleinere Schriften,* Vol. 1. Cotta, Stuttgart & Tübingen, 474pp.

Humboldt, A. von 1874a. *Kosmos, Entwurf einer physischen Weltbeschreibung,* 4 vols. Cotta, Stuttgart, 323, 348, 432, 536pp.

Humboldt, A. von 1874b. *Reise in die Aequinoctial-Gegenden des neuen Continents,* 2 vols. Cotta, Stuttgart 257 and 259pp.

Instituto Geográfico Militar 1974. *Ecuador, escala 1:1,000,000.* IGM, Quito.

Instituto Nacional de Meteorología e Hidrología, Ecuador 1974. Series climatológicas. Publicación No. 16-I, Quito.

International Association of Hydrological Sciences – UNESCO 1977. *Fluctuations of glaciers, 1970-75.* Paris, 269pp.

Jameson, W. 1861. Journey from Quito to Cayambe. *J. Roy. Geogr. Soc. 31:*184-190.

Johnson, A.M. 1976. The climate of Peru, Bolivia and Ecuador, pp. 147-218. In: W. Schwerdtfeger, ed., *Climates of Central and South America* (World Survey of Climatology, Vol. 12). Elsevier, Amsterdam-Oxford-New York, 532pp.

Juan, J. & A. de Ulloa 1748. *Relacion historica del viage a la America Meridional hecho de orden de S. Mag. para medir algunos grados de meridiano.* Part 1, Vol. 2, Madrid, 682pp.

Karsten, H. 1886. *Géologie de l'ancienne Colombie Bolívarienne, Vénézuela, Nouvelle-Grenade et Ecuador.* Friedlander & Sohn, 62pp.

Kennerley, J.B. 1971. *Geology of the Llanganates area, Ecuador.* Institute of Geological Sciences, Overseas Division, Photogeological Unit. London, Report No. 21, 13pp.

Kennerley, J.B. & R.J. Bromely 1971. *Geology and geomorphology of the Llanganati Mountains, Ecuador.* Instituto Ecuatoriano de Ciencias Naturales, Quito. Contribución No. 73, 10pp.

Kinzl, H. 1949. Die Vergletscherung in der Südhälfte der Cordillera Blanca. Begleitworte zu einer stereophotogrammetrischen Karte 1:100,000. *Zeitschrift für Gletscherkunde und Glazialgeologie 1:*1-28.

Kinzl, H. 1968. La glaciación actual y pleistocénica en los Andes Centrales. Proceedings of UNESCO-Symposium on Geo-Ecology of the mountainous regions of the Tropical Americas. *Colloquium Geographicum 9:*77-90.

Kinzl, H. 1970. Gründung eines glaziologischen Institutes in Peru. *Zeitschirft für Gletscherkunde und Glazialgeologie 6:*245-246.

Kolberg, J. 1885. *Nach Ecuador.* Herder, Freiburg, 550pp.

Koopmans, B.N. & P.H. Stauffer 1968. Glacial phenomena on Mount Kinabalu, Sabah. *Malaysian Geol. Survey, Bull. 8:*25-35.

La Condamine, C.M. de 1751. *Journal du voyage fait par ordre du Roi a l'Équateur, servant d'introduction historique a la mesure des trois premiers degrés du meridien.* Imprimerie Royale, Paris, 266pp.

Lewis, G.E. 1956. Andean geologic province, pp. 269-291. In: W.F. Jenks, ed., *Hand-*

book of South American geology. Geol. Soc. Amer. Mem. 65, 378pp.

Livingstone, D.A. 1962. Age of deglaciation in the Ruwenzori range, Uganda. *Nature* *194:*859-860.

Lliboutry, L., B. Morales Arnao & B. Schneider 1977. Glaciological problems set by the control of dangerous lakes in Cordillera Blanca, Peru. III. Study of moraines and mass balance at Safuna. *J. Glaciol. 18:*275-290.

Loeffler, E. 1972. Pleistocene glaciation in Papua and New Guinea. *Zeitschrift für Geomorphologie,* suppl. vol. 13, 32-58.

Loeffler, E. 1976. Potassium-Argon dates and pre-Würm glaciations of Mount Giluwe volcano, Papua, New Guinea. *Zeitschrift für Gletscherkunde und Glazialgeologie 12:*55-62.

Loeffler, E. 1979. Pleistocene and present-day glaciations, In: J.L. Gressit, ed., *Biogeography and ecology of New Guinea.* (Monographiae Biologicae) Junk, The Hague, in press.

Loor, W. ed. 1953. *Cartas de Gabriel Garcia Moreno 1846-1875,* 4 vols. La Prensa Católica, Quito, 326, 388, 424, 548pp.

Lorenzo, J.L. 1964. *Los glaciares de Mexico.* Monografías del Instituto de Geofísica, UNAM, No. 1, 124pp.

Martínez, N.G. 1933, *Exploraciones en los Andes Ecuatorianos, El. Tungurahua.* Imprenta Nacional, Quito, 117pp.

Mercer, J.H. 1967. *Southern hemisphere glacier atlas.* U.S. Army Natick Earth Sciences Laboratory, American Geographical Society, Technical Report 67-76-ES, 293pp + maps.

Mercer, J.H., L.G. Thompson, C. Marangunič & J. Ricker 1975. Peru's Quelccaya Ice Cap: glaciological and geological studies 1974. *Antarctic J. of the U.S. 10:*19-24.

Mercer, J.H. & O.M. Palacios 1977. Radiocarbon dating of the last glaciation in Peru. *Geology 5:*600-604.

Meyer, H. 1904a. Die gegenwärtigen Schnee- und Eisverhältnisse in den Anden von Ecuador. *Globus 85(10):*149-157.

Meyer, H. 1904b. Eiszeitliche Untersuchungen in den Anden von Ecuador, pp. 257-268, In: *F. Ratzel-Gedächtnisschrift.* Leipzig. .

Meyer, H. 1904c. Die Eiszeit in den Tropen. *Geographische Zeitschrift 10:*593-600.

Meyer, H. 1906-07. Der Calderagletscher des Cerro Altar in Ecuador. *Zeitschrift für Gletscherkunde 1:*139-148.

Meyer, H. 1907. *In den Hochanden von Ecuador,* 2 vols. Dietrich Reimer-Ernst Vohsen, Berlin. Vol. 1, 522pp, Vol. 2, picture atlas.

Moore, R.T. 1930. First ascent of Mt. Sangai, Ecuador. *Amer. Alpine J. 1:*228-229.

Morales Arnao, B. 1969. Las lagunas y glaciares de la Cordillera Blancy y su control. *Revista Peruana de Andinismo y Glaciología (1966-68) 8:*76-79.

Nogami, M. 1976. Altitude of the modern snowline and pleistocene snowline in the Andes. *Tokyo Metropolitan University Geographical Reports 11:*71-86.

Oppenheim, W. & H. Spann 1946. *Investigaciones glaciológicas en el Peru 1944-45.* Instituto Geológico del Peru, Bol. No. 5, 40pp.

Orton, J. 1868. Note on the physical geography of the Andes of Quito. *American J. of Science and Arts 45:*99-105.

Orton, J. 1869. Geological notes on the Andes of Ecuador. *American J. of Science and Arts 47:*242-251.

Orton, J. 1870. *The Andes and the Amazon.* Harper, New York, 356pp.

Osmaston, H.A. 1965. The past and present climate and vegetation of Ruwenzori and its neighborhood. Thesis, University of Oxford, 238pp.

Petersen, U. 1958. Structure and uplift of the Andes of Peru, Bolivia, Chile and adjacent Argentina. *Bol. Soc. Geol. del Peru 33:*57-129.

Petersen, U. 1967. El glacier Yanasinga, 19 años de observaciones instrumentales. *Bol. Soc. Geol. del Peru 40:*91-97.

Peterson, J.A. & G. Hope 1972. Lower limit and maximum age for the last major advance of the Carstensz glacier, West Irian. *Nature 240:*36-37.

Pflücker, L. 1905. *Informe sobre los yacimientos auriferos de Sandia.* Boletín del Cuerpo de Ingenieros de Minas del Peru, Lima, No. 26, 40pp.

Raasveldt, H.C. 1957. Las glaciaciones de la Sierra Nevada de Santa Marta. *Revista de la Academia Colombiana de Ciencas exactas, físicas, y matemáticas 9:*469-482.

Reiss, W. 1873. Ueber eine Reise nach den Gebirgen des Iliniza und Corazón und im Besonderen über eine Besteigung des Cotopaxi. *Zeitschrift der Deutschen Geologischen Gesellschaft 25:*71-95.

Reiss, W. 1875. Bericht über eine Reise nach dem Quilotoa und dem Cerro Hermoso in den ecuadorianischen Cordilleren. *Zeitschrift der Deutschen Geologischen Gesellschaft 27:*274-294.

Reiss, W. & A. Stübel 1892-98. *Reisen in Südamerika. Das Hochgebirge der Republik Ecuador. Petrographische Untersuchungen,* Vol. 1: *West-Cordillere;* Vol. 2: *Ost-Cordillere.* Asher & Co., Berlin, 358 and 356pp.

Rodriguez de Aguayo, Pedro 1965. Descripción de la ciudad de Quito y vecindad de ella, por el arcediano de su iglesia, Licenciado Pedro Rodriguez de Aguayo (Archivo de Indias), In: M. Jimenez de la Espada, *Relaciones geográficas de Indias – Peru,* 3 vols. Biblioteca de autores Españoles, Madrid, 415, 343, 281pp.

Sadler, J.C. 1975. The upper troposphere circulation over the global tropics. Department of Meteorology, University of Hawaii, UHMET-75-05, 35pp.

Sauer, W. 1950. Contribuciones para el conocimiento del cuaternario en El Ecuador. *Anales de la Universidad Central, Quito, 77(328):*327-364.

Sauer, W. 1957. *El mapa geéologico del Ecuador.* Editorial Universitaria, Quito, 70pp.

Sauer, W. 1965. *Geología del Ecuador.* Editorial de Ministerio de Educación, Quito, 383pp.

Sauer, W. 1971. *Geologie von Ecuador.* (Beiträge zur Regionalen Geologie der Erde, Vol. 11). Gebrueder Borntraeger, Berlin-Stuttgart, 316pp.

Servicio Nacional de Geología y Minería, Ministerio de Industrias y Comercio (1969). *Mapa geológico de la República del Ecuador.* IGM, Quito.

Servicio Nacional de Meteorología e Hidrología, Ecuador 1971. *Mapa pluviométrico parcial del Ecuador.* (Valores normales 1939-60.) Quito.

Servico Nacional de Meteorología e Hidrología Ecuador 1972. *Boletín climatológica, Quito,* Año 11, 1972.

Schottelius, J.W. 1935-36. Die Gründung Quitos. Planung und Aufbau einer spanischen Kolonialstadt. *Ibero-Amerikanisches Archiv, 9:*159-182, 276-294; *10:*55-77.

Schubert, C. 1972a. Geomorphology and glacier retreat in the Pico Bolivar area, Sierra Nevada de Merida, Venezuela. *Zeitschrift für Gletscherkunde und Glazialgeologie 8:*189-202.

Schubert, C. 1972b. Late glacial chronology in the Northeastern Venezuelan Andes. *24th International Geol. Congr. Montreal, Sect. 12,* pp. 103-109. Montreal.

Schubert, C. 1974. Late pleistocene Mérida glaciation, Venezuelan Andes. *Boreas 3:* 147-152.

Schubert, C. 1975. Glaciation and periglacial morphology in the Northwestern Venezuelan Andes. *Eiszeitalter und Gegenwart 26:*196-211.

Schunke, E. 1975. *Die Periglazialerscheinungen Islands in Abhängigkeit von Klima und Substrat.* (Abhandl. Akademie der Wissenschaften in Göttingen.) Vandenhoeck-Rupprecht, Göttingen, 272pp.

Sievers, W. 1908. Vergletscherung der Cordilleren des tropischen Südamerika. *Zeitschrift für Gletscherkunde 2:*271-284.

152

Sievers, W. 1914. *Reise in Peru und Ecuador, ausgeführt 1909* (Wissenschaftliche Veröffentlichungen der Gesellschaft für Erdkunde zu Leipzig, Vol. 8.) Dunker & Humlot, Munchen-Leipzig, 411pp.

Spruce, R. 1861. On the mountains of Llanganati in the Eastern Cordillera of the Quitonian Andes. *J. Roy. Geogr. Soc. 31;* 163-184.

Stübel, A. 1886, *Skizzen aus Ecuador.* Asher & Co., Berlin, 96pp.

Stübel, A. 1897. *Die Vulkanberge von Ecuador.* Asher & Co., Berlin, 557pp.

Temporary Technical Scretariat for World Glacier Inventory of UNESCO-UNEP-IUGG-IASH-ICSI 1977. *Instructions for compilation and assemblage of data for a World Glacier Inventory.* ETH, Zürich, 29pp.

Thompson, L., Hastenrath, S., Morales & B. Arnao 1979. Climatic ice core records from the tropical Quelccaya Ice Cap. *Science, 203:* 1240-1243.

Tschopp, H.J. 1948. Geologische Skizze von Ekuador. *Bull. Ver. Schweiz. Petroleumgeol. und Ing. 15(48):* 14-45.

UNESCO 1970. *Perennial ice and snow masses. A guide for compilation and assemblage of data for a World inventory.* Technical Papers in Hydrology, No. 1.

U.S. Weather Bureau, ESSA, NOAA 1965-76. Monthly climatic data for the World, years 1965-75. Asheville, N.C.

Van der Hammen, Th. 1974. The pleistocene changes of vegetation and climate in tropical South America. *J. Biogeography 1:* 3-26.

Velasco, J. de 1841-44. *Historia del reino de Quito,* 3 vols. Imprenta del Gobierno, Quito, 287, 210, 252pp.

Villavicencio, M. 1858. *Geografía de la Republica del Ecuador.* Craighead, New York, 505pp.

Wagner, M. 1870. *Naturwissenschaftliche Reisen im tropischen Amerika.* Cotta, Stuttgart, 632pp.

Wisse, S. 1849. Études sur les blocs erratiques des Andes de Quito. *Comptes Rendues de l'Academie des Sciences, Paris, 28:* 303-307.

Wisse, S. & G. García Moreno 1846. Exploration du volcan Rucu-Pichincha, pendant le mois d'août 1845. *Comptes Rendues de l'Academie des Sciences, Paris, 23:* 26-35.

Whittow, J.B., J.E. Goldthorpe & P.H. Temple 1963. Observations on the glaciers of the Ruwenzori. *J. Glaciol. 4:* 581-616.

Whymper, E. 1892. *Travels amongst the Great Andes of the Equator.* Charles Scribner's Sons, New York, 456pp.

Wolf, T. 1879. *Viajes científicos por la República del Ecuador,* 3 vols. Guayaquil, 57, 78, 87pp.

Wolf, T. 1892. *Geografía y geología del Ecuador.* Brockhaus, Leipzig, 671pp.

Wood, W.A. 1970. Recent glacier fluctuations in the Sierra Nevada de Santa Marta. *Geogr. Rev. 60:* 374-392.

World Meteorological Organization – ICSU 1975. *The physical basis of climate and climate modelling.* GARP Publication Series, No. 16, 265pp.

AUTHOR INDEX

.

SUBJECT INDEX

157

159

T - #0654 - 071024 - C200 - 234/156/10 [12] - CB - 9789061910381 - Gloss Lamination